ENVIRONMENTAL POLLUTION : CAUSES, EFFECTS AND CONTROL

Environmental Pollution: Causes, Effects & Control

Inderjeet Sethi
Department of Botany
Khalsa College
(University of Delhi)
Delhi-110007

M. S. Sethi
Department of Chemistry
A.R.S.D. College
(University of Delhi)
New Delhi-110021

S. A. Iqbal
Department of Chemistry
Saifia P.G. College of Science & Education
Bhopal-462001

Commonwealth

Environmental Pollution : Causes, Effects & Control

ISBN : 81-7169-123-4

Edition : 2026

Published by :
COMMONWEALTH PUBLISHERS (P) LTD.
4831/24, Govind Lane,
Ansari Road, Darya Ganj,
New Delhi - 110002
Ph. : 23272541, 23242541
e-mail : commonwealthpublishers2016@gmail.com

Distributed by :
ARJUN PUBLISHING HOUSE
4831/24, Govind Lane,
Ansari Road, Darya Ganj,
New Delhi - 110002
Ph. : 23272541, 23242541
e-mail : arjunbooks2001@gmail.com

Laser Typesetting by :
Dimple Computers
Delhi

Printed at :
Balaji Offset
Delhi

Preface

The environment is a variegated pattern of biological and physical components. The relationship between living organisms and their natural environment remains as important as ever, but more emphasis is now being placed on the disruptive effects of man's activities. Environmental pollution has been there for a long time; today we are all aware of the harmful effects of pollution on human health and our environment as a whole. The effects and the extent of environmental pollution are a matter of serious concern for the society, both now and in the future.

It is universally recognised now that pollution is an increasing environmental problem that needs serious objective study. This requires factual information relating to waste production and disposal, the levels of pollution and the effects of pollutants on our environment. This information is vital for the assessment of pollution problems and control.

Pollution is an integrated phenomena, that must be considered within the natural systems that exist in the environment. However, in order to study pollution, it is convenient to subdivide it into artificial segments of atmosphere, land, water and noise pollution. Environmental pollution problems involve understanding of elementary knowledge of biology, chemistry and physical environmental phenomena. Therefore, we have included sufficient scientific background in this book to enable the readers to understand the problems of pollution.

Our theme in this book is that people must learn to live with nature and not destroy it. We must look at solutions to our environmental problems. These solutions come from an understanding of how we interact with nature. The most practical solutions we can invoke to solve our environmental problems are those which involve recycling, using alternative sources of energy, significantly reducing all kinds of pollution in our ecosystems, educating people how best to use and manage our resources, and starting to be responsible stewards of our natural environment. These solutions obviously mean altered life-styles. The time is now at hand when we must stop exploiting nature for our own appetites and start living with nature by returning to it the materials we use, by keeping our populations under control and by following the ways of nature (*e.g.*, by using composting, biogas generators, fusion, wind power, and solar power for energy needs). Otherwise, this society, unless it changes its collision course with nature, will perish from the face of the earth.

The authors are grateful for the help and suggestions provided by a cross-section of people—research workers, scientists and authors. We wish to express our appreciation for the help and suggestions given by Dr. Gurcharan Singh, Reader in Botany, Khalsa College, University of Delhi, Delhi and Dr. Subodh K. Sharma, Scientist-F, DOE, Ministry of Environment and Forests, Government of India, New Delhi.

AUTHORS

Contents

PART II

THE QUALITY OF ENVIRONMENT: CAUSES AND EFFECTS OF ENVIRONMENTAL POLLUTION

PART III

POLLUTION CONTROL

1

Introduction

The word "Environment" encompasses everything that is around us. To the layman it means rather all things that can be damaged by man's activities. The real impact of Man on the environment began when mining acquired a definite importance, that is only 40 to 50 centuries ago. And environmental effects took a significant importance only since the explosive industrial development began, around 1780, starting the exponential growth of world population and consumption.

The first environmental laws were enacted nearly at the same time. Napoleon ordered in 1810 that measures be taken by all factories to prevent damage or any kind of trouble to neighbours, crops, buildings, etc. Similar regulations were soon imposed in all industrial countries. Nevertheless it was only during the past thirty years that our world has really become "environment-conscious". The wide-ranging character of pollution, of the irreversible damage it can cause in some cases, is a very recent stage of our conscious knowledge. And because it is still very young, the field of environmental protection remains incompletely defined in a number of domains.

A major part of the pollutants can be carried by air, sometimes over long distances, which can eventually reach the ground and be turned into soil and water pollution, reach the troposphere and stratosphere and be a grave cause for concern (in particular the possible depletion of the protective ozone layer), and also reach our eyes and lungs through smog or "ordinary" pollution. A quite unhealthy picture, which gives a poor idea of the way Man, the Lord of Creation, has dealt with his environment and created the conditions of his own discomfort or hazard. This hazard can sometimes be frightening: the memories of Bhopal are in all the minds.

The sources of pollution are many, but the general public thinks chiefly of three risks: (a) Oil spills, the most recent one has occured in the Persian Gulf on 26th January 1991 during the Gulf War between Iraq and U.S. led multinational forces. The oil slick is reported to be 140 km long and 40 km wide and is spreading southwards at the time of writing. (b) Nuclear hazards, because at Hiroshima and Nagasaki. The layman does not make the difference between a drug and a poison, between "the" bomb and a nuclear reactor, although the nuclear industry is abundantly proved to be the safest of all industries; (c) The chemical industry, the permanent villain, judged to be the cause of all pollution, the source of bad smells and a danger to all.

Chemistry and the chemical industry have not been able to give of themselves a faithful image. It may bear the responsibility for some damage to our world, and we all have to be informed of the part of chemistry in the Bhopal tragedy. But chemistry is a science of creation, the chemical industry brings to mankind priceless contributions: drugs, pesticides, fertilizers, synthetic textiles, etc. and both chemical research and the chemical industry will probably be in charge of finding the proper solutions to the environmental problems linked to the technical developments of the contemporary world.

The chemists have to face a number of difficulties in their relationship to the public, a by-product of the emission of many pollutants by the chemical and related industries. Despite all that comes from other sources or causes, what comes to mind when chemistry is mentioned is bad smells, pollution, toxicity. The benefits we all draw from chemistry in our well-being, comfort, health, safety and economy can be priceless, the layman will in most cases look at us and say "Bhopal". It may be unfair, that the assets of chemistry are just taken for granted, but life is often unfair. The Bhopal tragedy is the worst industrial disaster probably known. What happened was the result of an important leak of gaseous methyl isocyanate (MIC) from a number of equipments which were out of order due to the neglect of several safety requirements and operating instructions. It is very sad to think that most of the victims would have been saved if they had protected their face with a damp cloth, and that this so simple fact was ignored.

It seems to be true that industrial production increases pollution but things are fortunately not so simple and not so bad. Many of the industrial wastes are raw materials or end products, and preventing their escape is a saving at the same time as an environmental protection. Other pollutants are really refuse, but in all cases the fight against pollution is on. In fact the pollution level has appreciably decreased in all developed countries in recent years. It is probable that the public is not really aware of the fact, and sees in pollution what remains, not what has been removed. And although industrial pollution has really decreased, the fact remains that a number of problems are as yet unsolved, among which the damage to forestry and plants figures prominently.

For a number of years damage to vegetation, and most of all to forests. have been evident in several countries. In Europe the extent of damage is very grave in Germany, Czechoslovakia and Poland, slighter but not negligible in other countries. It seems quite certain that this damage is related to the presence of some of the best-known gaseous pollutants, sulphur dioxide, nitrogen oxides, ozone, and also perhaps hydrocarbons.

In the case of plant damage through air pollution no lives are at risk but our landscapes are menaced and the economic damage is very high. This is the type of problem we encounter in several cases. The origins of the forestry problem are far from being completely clear, so that even choice of remedies has to be very carefully weighed, Whatever the remedy the cost will be quite impressive. An analogous situation has been met in all cases of large-scale damage. Cleaning the Ganga water for instance, involve a huge investment. Several other examples could be given and, to mention one of the serious European problems, cleaning the Rhine will be an expensive task. In the case of the Rhine Switzerland, the Federal Republic of Germany, France and the Netherlands are all polluters. The heaviest damage is suffered by the Netherlands, for whom the Rhine is the only source of sweet water, and the heaviest single polluter are the Alsatian salt mines. All Mediterranean countries are damaging their sea, but it is evident that pollution comes chiefly from industrialized countries on the Northern shores. The USSR does probably more for the pollution of the Black Sea than Bulgaria etc.

Turning towards the sources of this pollution, we might think that the SO_2 emission from volcanoes cannot be checked but certainly from a power-station. However, the costs would be tremendous, whether the power-station burns coal or heavy fuel oil, both sulpur-rich, but emitting a plume in which SO_2 is quite diluted. The greater the dilution the higher the cost of recovery.

With nitrogen oxides the situation is comparable. Thunderstorms are great sources of nitrogen oxides but an abundant daily supply comes from car exhausts-among others. The problem has been studied now for a number of years. Car exhausts can be adapted and improved some day, at a higher cost of the car and its maintenance. Quite similar considerations would also apply in the case of other pollutants.

It is established that in many places the damage to plants continued to increase, whereas pollution had markedly decreased. That is the reason for the careful theories on the effects of accumulating pollutants and the mechanisms through which this effect is felt. Now nearly all theories take into account accumulation effects, but the parameters are many. Damage to forests has been shown to be altitude-dependent, raising the possibility, that the chief cause of damage might be the nitrogen oxides-ozone chemistry. Another view is the possibility of long-term interaction between soil nutrients and air pollutants having accumulated in the soil. It could also be argued that no single compound is responsible for the observed damage, because several countries are submitted to the same pollution level with the same pollutants but widely differing damage. The possible synergy between a well-known pollutant present everywhere at low concentrations, and a second, "activating" compound, which would trigger the synergy, cannot be ruled out.

Acid rains have certainly played a part in damage to trees but the existing theories are not in complete agreement. A new crop of facts, a new idea, and fresh hypotheses are coming up each day, all of which have to be considered.

Today we are faced with a grave and difficult situation and, in order to improve things, we need a better comprehension of the factors involved in environmental effects.

Our theme in this book is that people must learn to live with nature and not destroy it. We must look at solutions to our environmental problems. These solutions come from an understanding of how we interact with nature. The most practical solutions we can invoke to solve our environmental problems are those which involve recycling, using alternative sources of energy, significantly reducing all kinds of pollution in our ecosystems, educating people how best to use and manage our resources, and starting to be responsible stewards of our natural environment. These solutions obviously mean altered life-styles. The time is now at hand when we must stop exploiting nature for our own appetities and start living with nature by returning to it the materials we use, by keeping our populations under control and by following the ways of nature (e.g., by using composting, biogas generators, fusion, wind power, and solar power for energy needs). Otherwise, this society, unless it changes its collision course with nature, will perish from the face of the earth.

PART I

ENVIRONMENTAL SEGMENTS

2

The Atmosphere

The atmosphere is an insulating blanket of air around the earth. Without it the temperature at the equator would rise to 82°C during the day and drop as low as—140°C at night. It burns up meteors that would bombard the surface of the earth from space. The atmosphere does many things to make life possible on earth. It is a reservoir of oxygen needed by animals and of carbon dioxide essential for plants. It serves as a distribution system for moisture. Without the atmosphere there would be no sound and no flight. There would be no conventional long-distance radio communication, because this is dependent on the electrons in the upper atmosphere. Without air there would be no lightning, no clouds, no wind, no rain, no snow, and no fire. The surface of the earth would be as bleak and sterile as the moon. Fig. 2.1 illustrates an atmospheric model.

The atmosphere shields the earth from lethal concentrations of ultraviolet radiation. It selectively filters the sun's rays so that only two small segments of the electromagnetic spectrum penetrate to the earth's surface in appreciable amounts. One segment is the *optical window*, consisting essentially of the visible spectrum of light, from near-ultraviolet to near-infrared. The other is the *radio window*, consisting of radio waves from about 1 centimetre to 40 metres in length.

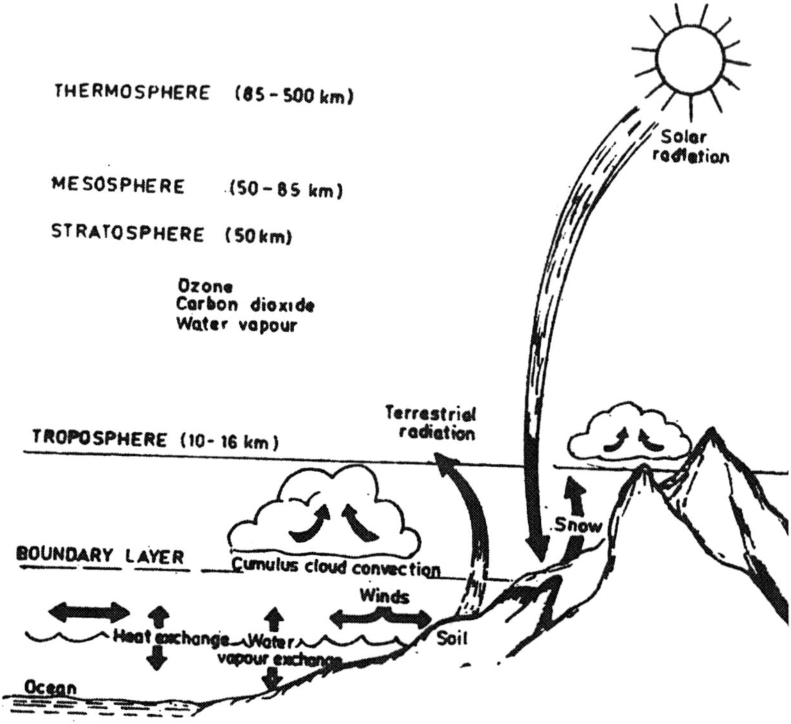

Fig. 2.1. An atmospheric model.

It is impossible to define the limits of the atmosphere because the atmosphere becomes progressively thin with increasing distance from the earth. There is no boundary between the atmosphere and the void of outer space. However, 75 per cent of the earth's atmosphere lies within 16 km of the surface, and 99 per cent of the atmosphere lies below an altitude of 30 km. The components of the atmosphere are listed in Table 2.1.

Clean, dry air at sea level contains, by volume, 78 per cent nitrogen gas, 21 per cent oxygen, 1 per cent argon, and 0.03 percent carbon dioxide, plus small amounts of other gases. The latter (in approximate order of decreasing concentration) include neon, helium, methane, krypton, sulphur dioxide, hydrogen, nitrous oxide, xenon, ozone, nitrogen dioxide, iodine, ammonia, and carbon monoxide (Table 2.2).

The nitrogen gas is biologically inert for most organisms, except under special conditions to be discussed later. It normally takes no part in the role of animal life except as a diluent for the oxygen and carbon dioxide, but it is ultilized by some bacteria and some plants. Oxygen is essential to the production of biological energy in the burning of cellular fuel. Carbon dioxide, despite its low concentration (less than 1/600 that of oxygen), plays a vital role in the cellular processes of animals and plants. Its most important, use is by plants in photosynthensis, by which the green pigment chlorophyll uses carbon dioxide, water, and sunlight to synthesize sugars and starches. In animals, carbon dioxide is a waste product, but it also regulates the rate of breathing. Carbon dioxide also performs an important function in controlling the acidity-alkalinity balance of the blood by its presence in the form of carbonic acid, identical to the product in carbonated water.

However, all of the major, and some of the minor, atmospheric components are biologically important. The participation of each in the living processes is in some cases critically sensitive to slight changes in concentration.

TABLE 2.1. **Major Regions of the Atmosphere**

Region	*Altitude range (km)*	*Temperature range (°C)*	*Significant chemical species*
Troposphere	0—10 to 16	15 to —56	N_2, O_2, CO_2, H_2O
Stratosphere	10 to 16—50	—56 to —2	O_3
Mesosphere	50—85	—2 to—92	O_2^+, NO^+
Thermosphere	85—500	—92 to 1200	O_2^+, O^+, NO^+

TABLE 2.2. **Composition of Clean, Dry Air at or Near Sea Level**

Component	*Percent by volume*
Nitrogen	78.084
Oxygen	20.9476
Argon	0.934
Carbon dioxide	0.0314
Neon	0.001818
Helium	0.000523
Methane	0.0002
Krypton	0.000114
Sulphur dioxide	0 to 0.0001
Hydrogen	0.00005
Nitrous oxide	0.00005
Xenon	0.0000087
Ozone	Summer 0 to 0.000007
	Winter 0 to 0.000002
Nitrogen dioxide	0.000002
Iodine	0 to 0.000001
Ammonia	0 to trace
Carbon monoxide	0.1 ppm—0.2 ppm

Major Regions of the Atmosphere

Troposphere

The zone nearest the earth, the troposphere, derives its name from the Greek *tropein* (to turn, to rotate, to change). One property that changes in the troposphere is the temperature. It drops to —55°C by the time an altitude of about 16—18 km is reached. The rate of temperature drop is about 6.4°C per 1000 metres. This is called the *normal environmental lapse rate*. The region through which this lapse rate occurs defines the troposphere. At the equator the troposphere is about 18 km thick, but at the poles it is only about 8 km thick. Essentially all the atmosphere's water vapour is in the troposphere, which therefore encloses essentially all the storms and precipitation of the earth. The jet stream occurs at an altitude of 10.5—12 km.

Stratosphere

This zone begins at the top of the troposphere, at a min atmospheric "shell" called the tropopause, where the temperature begins to stay fairly constant, increasing slightly, in going to an altitude of 50 km. The stratosphere includes much of the *ozone layer*. The warming trend in the stratosphere is caused by a cycle of chemical changes in the ozone layer which converts all the incoming high energy ultraviolet rays to heat. If these rays reached the earth's surface, life as we know it would be impossible. The *stratopause* marks the narrow zone at the top of the stratosphere where the temperature begins to fall again with increasing altitude as the mesosphere is entered.

Mesosphere

The mesosphere, on top of the stratosphere, extends roughly to 80 km and its temperature drops to between −80 and —90°C at the mesopause, a thin zone where the temperature stabilizes and soon starts to rise again as the *thermosphere* is ascended.

Thermosphere

This zone is above 80 km altitude. The temperature becomes very high, but the air has such a low density that it can hold very little heat. Any denser object in the thermosphere will be extremely hot in sunlight but very cold at night.

The zones just discussed are defined by changes of temperature with altitude. Changes in chemical composition or activity also occur with altitude. Up to about 80 km the proportions of components given in Table 2.2 remain nearly constant, and the long zone up to that altitude is sometimes called the *homosphere*. Within it is a region called the *chemosphere* extending from the upper troposphere where a huge number of chemical changes occur powered by solar radiation. Ozone production is one such activity. Above the homosphere lies the *hetrosphere*. For its first 40 km, the *ionosphere*, the very thin concentration of atoms and molecules, exists largely as electrically charged particles called ions. This globe-enveloping band of ions reflects outgoing radio waves back to earth, making radio transmission possible well beyond the horizon. Television waves are not reflected, and a TV transmitter's range does not extend beyond the horizon. Meteorites entering the earth's atmosphere generally burn out in the ionosphere.

Radiant Energy

Heat for the atmosphere comes from the *radiant energy* given off by the sun. The sun's energy reaches, the earth as sunlight. Light waves are absorbed by the earth and returned to the air as heat waves. However, only a part of the sun's radiant energy reaches the earth. Some energy is reflected into space by gas molecules, by clouds, and by the earth itself. Some of the energy is absorbed by the ozone layer. Of the radiation that reaches the atmosphere from the sun, only 30 to 80 per cent reaches the earth's surface (Fig. 2.2). The amount that reaches Earth depends largely on cloud cover and altitude. Areas near the equator receive the most energy because the air envelope is thinnest. The sun's rays reach the earth at nearly a 90 degree angle at the equator. At the poles, the sun's rays strike the earth at about a zero degree angle. Energy received by the earth decreases from the equator to the poles.

Earth's surface features influence the amount of energy absorbed by the earth. Polar ice reflects the sun's rays, and little energy is retained. Oceans also reflect much of the sun's energy when the sun is close to the horizon. Sand and snow

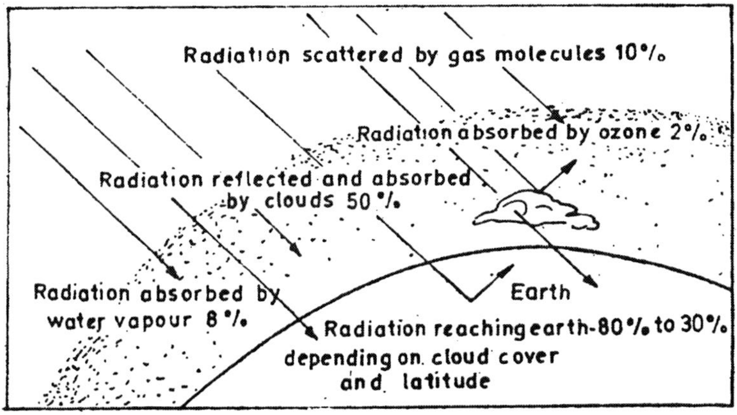

Fig. 2.2. A varying per cent of the sun's radiant energy reaches the earth through the protective blanket of the atmosphere.

are other good reflectors. Vegetation, dark rocks, and asphalt tend to absorb the sun's rays.

Most of the sun's radiant energy that reaches the earth comes from the visible light spectrum. Somc sunlight is used by green plants in their food-making process. Some sunlight is reflected by the earth. But light rays that are absorbed by Earth are changed to heat waves. These heat waves are the major source of heat for the atmosphere.

When the sun sets, Earth and atmosphere lose heat to outer space. Land areas lose heat more quickly than oceans do. Deserts and mountaintops also lose heat rapidly. Earth would have great differences in day and night temperatures if it were not for the atmosphere.

Both carbon dioxide and water vapour absorb the heat given off by Earth. This heat then is returned to Earth. The effect is like a greenhouse. In the greenhouse, sunlight comes in through the windows. Heat, however, is prevented from escaping because of the screening effect of the glass. Heat loss to outer space is least wherever large amounts of water vapour are in the air.

Air Circulation

Air in the troposphere is constantly in motion. Horizontal movements of air are called *winds*. Vertical movements of air are called *currents*. Winds and currents together carry heat from the equator to the poles. This heat comes directly from the earth but indirectly from the sun. Winds and currents determine the major circulation pattern of the air as they move air from high pressure areas to low pressure areas (Fig. 2.3). High pressure is associated with cold air. Low pressure generally is associated with warm air. The density of air is determined by its temperature and the amount of water vapour per unit volume. A given volume of water vapour has less mass than the gases it displaces. Thus, moist air is less dense than dry air.

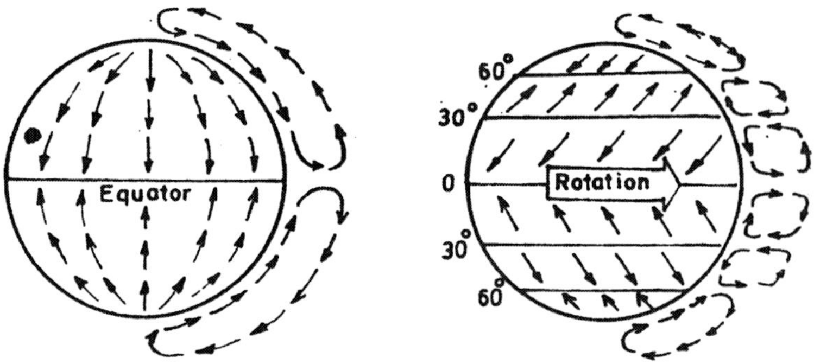

Fig. 2.3, Major north-south convection currents in the atmosphere are broken into eddies by the earth's rotation, forming the wind patterns of the globe.

The portion of the earth between latitudes 23.5° north and south of the equator receives the most radiant energy. Therefore, air at the equator is warmer and less dense than air at other latitudes. Cool surface air flows toward the equator and forces currents of warm air to rise. In the upper atmosphere, warm air flows toward the poles. This air gradually cools and at the poles it is cold, dry air. Cold air sinks to the earth's surface in the polar region. From the poles, the air flows toward the equator. This circulation of air carries heat from the equatorial region to the rest of the earth. But air circulation is not a simple north-south movement. Circulation is complicated because Earth rotates on its axis and areas of land and sea are unequal.

Coriolis Force

Coriolis force is due to differences in velocity of rotation of the earth. At the equator, the velocity of earth's rotation is 1600 km per hour. This is the maximum rate of rotation. At the poles, the velocity of rotation is zero, the minimum rate. Velocity of rotation gradually decreases from the equator to the poles.

Circulation of air caused by convection alone is in a north-south direction. But the earth turns toward the east, and this causes winds to appear to curve away from a straight course. For example, a particle of air that moves directly *north* appears to be deflected toward the west in the Southern Hemisphere. A particle of air that moves directly *south* appears to be deflected toward the west in the Northern Hemisphere. When the air particle reaches the equator, the point toward which it was moving already has moved some distance eastward. Thus, the direction of travel appears to be from northeast to southwest in the Northern Hemisphere; from southeast to northwest in the Southern Hemisphere.

Major Wind Systems

The Doldrums

The doldrums is a windless zone at the equator. There, air seems to be motionless. Actually, it is rising. Sailing vessels often were stranded in the doldrums for many days because there were no winds to move them. Fig. 2.4 illustrates the major wind systems.

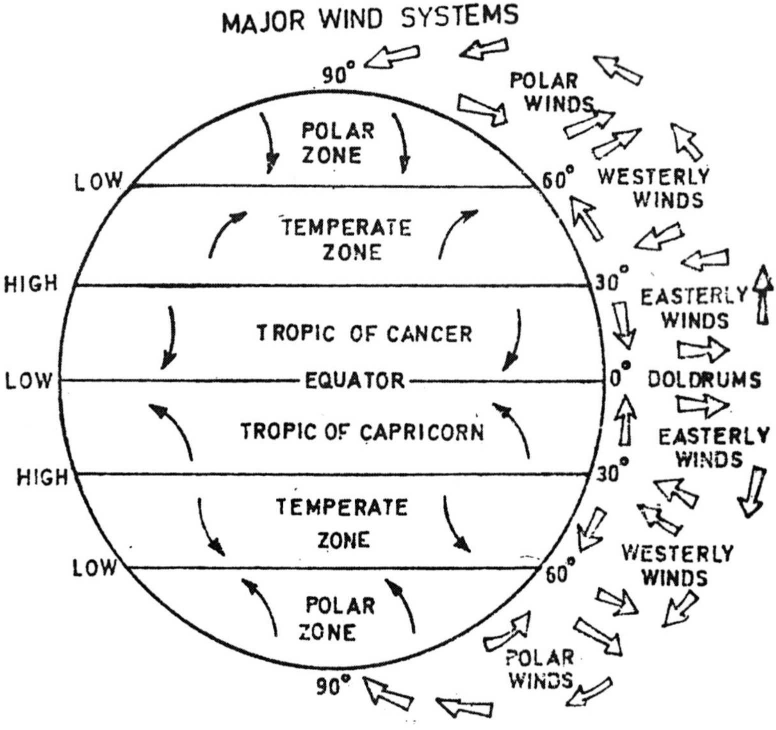

Fig. 2.4, Surface winds blow from regions of high density descending currents toward regions of low-density ascending currents.

The Easterlies

Trade winds blow from about the 30° latitude to the doldrums in both Northern and Southern Hemispheres. Trade winds blow northeast to southwest in the Northern Hemisphere. They blow from southeast to northwest in the Southern Hemisphere. Since winds are named for the direction from which they blow, the trade winds are called *easterlies*. These easterlies provided a popular sailing route for early trading ships. Today, airplane pilots take advantage of the easterlies when traveling toward the west in the tropics.

The Westerlies

The *westerlies'* zone lies between 30° and 60° latitude. At 30° latitude, air currents descend to the earth's surface. Part of this air turns southward as the trade winds. Part of it flows northward as the westerlies. Winds blowing toward the poles are deflected toward the east because of the Coriolis force. Westerlies are the air currents responsible for the movement of weather across the United States and Canada.

The Polar Winds

Polar wind zones lie between the poles and 60° latitude. Upper air currents flow poleward from the equator. These currents are cold, dry, and dense by the time they descend to the earth's surface near the poles. From the poles, air flows toward warmer latitudes at the earth's surface. Although the polar zones are relatively small, the winds contribute large amounts of energy to the westerlies.

Oxygen

Oxygen Content of the Atmosphere

No change in oxygen content of the air can be detected in records since 1910. According to calculations by Broecker (1970) each square metre of the earth's surface is covered by 6×10^4 moles (equal to more than a ton) of oxygen gas. Terrestrial, aquatic, and marine plants produce about 8 moles of oxygen gas annually for each square metre of the earth's surface. However, nearly all of this product of photosynthesis is utilized by animals and bacteria with the result that the amount that they consume is practically equal to the oxygen released into the atmosphere by plants.

There is a small net addition of oxygen to the atmosphere equal to about 1 part in 15 million parts of the oxygen pre sent. Probably much of the excess is removed by oxidation of carbon, iron, sulphur, and other minerals during the normal process of weathering. Unless cataclysmic* changes occur, the oxygen content of the atmosphere is not susceptible to significant change over a geologically short-range time scale of 100 to 1,000 years.

A major function of the blood is to carry oxygen from the point of entry at the alveolar membranes to the tissues of the body. The blood plasma contains 2 to 3 per cent dissolved oxygen . But its capacity is limited, thus animals that have to depend on the plasma for oxygen transport must be small or have other mechanisms for transferring oxygen and carbon dioxide within their bodies. In higher animals most of the oxygen is carried in the red blood cells, which are essentially bags of haemoglobin.

An adult at rest and breathing quietly ordinarily takes into the lungs about $^1/_2$ litre of air with each breath. At a regular respiration rate of 15 or 16 times a minute, a man (it is higher for women) will inhale and expire $7^1/_2$ to 8 litres of air a minute, or 475 litres an hour. During strenuous exercise, the higher oxygen demand causes an increase in both the rate and depth of respiration, so that 10 to 20 times as much air will be used in a given period.

The air, like all substances, has weight and is prevented from flying off into space by the pull of the earth's gravity. It is evident that at high altitudes the lungs have to 'work' harder because the concentration of oxygen is lower and there is less diffusion pressure to force the molecules through the membrane.

Oxygen Toxicity

Too much oxygen, called *hyperoxia*, is as dangerous as oxygen deficiency. Pure oxygen at atmospheric pressure can be

* A cataclysmic evenr is one that changes a situation or society greatly, especially in an unpleasant manner.

tolerated without harm for about 24 hours by the average person. After that time, symptoms of irritation appear followed by fluid in the lungs and other discomforting symptoms. Hospitals often administer oxygen continuously for days or weeks, for which purpose 60 per cent oxygen is considered to be the maximum safe concentration, and even lower concentrations are sometimes needed to avoid danger.

Scuba and deep-sea divers would have a similar problem if they were to breathe pure oxygen, because the pressure at a depth of 10 metres is twice that at the surface and increases one atm. with every increase 10 of metres. At a depth of 30 metres the pressure is four times that at the surface. A diver breathing pure oxygen will eventually suffer oxygen toxicity. Therefore, compressed air is generally used, although mixtures of oxygen and gases other than nitrogen, such as helium, have been used for deep diving.

The effect of too much oxygen resembles the toxic effect of the powerful poison strychnine. There is a profound effect on the muscular system and the entire central nervous system. The victim is liable to have headache and nausea recurring perioidcally, muscular twitehing, dizziness, impaired vision, irritability, and possibly convulsions that resemble epileptic seizures. Headache, drowsiness, and dizziness may persist for several hours.

Nitrogen

Nitrogen gas is the most abundant component of the atmosphere, comprising 78 per cent by volume. In some bacteria and blue-green algae nitrogen fixation is an important means of augmenting the nutrient supply, especially for higher plants when they grow in association with the nitrogen-fixing microorganisms.

Although nitrogen comprises the major bulk of the air, it is almost inert insofar as animals are concerned, except when one breathes air under increased pressure, such as when scuba or deep-sea diving, at which time the effects of this gas can be catastrophic. Nitrogen is a deadly threat to divers, who may fall victim to absorption of the gas into the blood stream. This affliction is known to divers as the "*bends,*" or *Caisson's disease*.

The high partial pressure of nitrogen during dives forces the gas readily through the alveolar membranes. Under a pressure sometimes several times the normal atmosphere pressure, depending on depth, nitrogen diffuses rapidly into the blood stream. But the tissues have no mechanism for chemically absorbing it as with oxygen and carbon dioxide. Nitrogen can be eliminated only through the lungs. Under high pressure it remains in solution in the plasma and during deep dives the nitrogen in the blood is thus greatly increased. If ascent is too rapid, diffusion through the alveolar membrane does not take place quickly enough to allow the nitrogen to leave the blood and return to the air in the lungs. Instead, the nitrogen may bubble out in the blood vessels like gas from a bottle of soda water when the pressure is suddenly released by removing the cap. The "bends" is a serious and sometimes fatal condition. It can be prevented by rigidly observing the rules for slow return to the surface, for which instruction and practice are required.

Nitrogen Narcosis

Breathing nitrogen under pressure has an intoxicating effect. At depths of about 10 metres and below, sufficient to increase the pressure to 4 atmospheres or more, the amount of nitrogen forced into the blood stream through the alveolar membrane reaches a concentration at which the nitrogen has a narcotic or anaesthetic action. It is a delightful but dangerous sensation. The condition is called *nitrogen narcosis* or "rapture of the deep."

Nitrogen narcosis is similar in effect to alcohol intoxication. The diver becomes uncoordinated and lacks judgment. The most dangerous part is that under the influence of nitrogen narcosis, the diver may want to go deeper, which could be fatal. Nitrogen narcosis becomes so severe at a depth of 300 feet that a man may be unable to perform even simple tasks. But the affected diver recovers quickly upon return to the surface.

Nitrogen Fixation

All living organisms need nitrogen in large amounts. Nitrogen is a key element in the amino acids as well as in many other dynamic molecules of the protoplasm. The use of nitrogen compounds in the metabolic manufacture of nutrients within the tissues and the fate of nitrogen products during and after decomposition are part of a complicated series of processes referred to as the *nitrogen cycle* (See Chapter 5). Much of the nitrogen in animals, plants, and their waste products is returned to the soil and reused by plants as part of the cycle. However, microorganisms called *denitrifying bacteria* act on soil nitrates and nitrites and in doing so produce molecular nitrogen (nitrogen gas), which returns to the atmosphere. Probably all of the nitrogen available for food in the soil would be lost if it were not for another part of the nitrogen cycle in which atmospheric nitrogen is eventually returned to the soil.

Nitrogen gas is a chemical substance of moderate chemical stability and does not readily enter into chemical reactions. But its abundance in the atmosphere provides an almost unlimited source of nitrogen. In nature, there are primarily two ways in which nitrogen is returned to the soil. One is by the energy of lightning which forms small amounts of nitrogen oxides from molecular nitrogen. The nitrogen oxides drop to the earth in moisture and become available for further modification by microorganisms as plant nutrients. A more important way in which nitrogen becomes "fixed", is by the action of nitrogen-fixing organisms among the algae and bacteria. It is one of the most crucial links in the nitrogen cycle.

Two groups of bacteria are largely responsible for nitrogen

fixation: (1) those which live in the soil independently of higher plants, and (2) those which grow in association with higher plants and have a mutual relationship with them. The relationship of two species of organisms living together in an intimate association is called *symbiosis*. The particular kind of symbiosis in which different kinds of organisms living together are of benefit to each other is called *mutualism*.

There are two types of nonsymbiotic nitrogen-fixing bacteria, those which require oxygen and those which do not. *Aerobic* (oxygen-requiring) bacteria called *Azotobacter* do the work in alkaline soil, while *anaerobic* bacteria with the formidable name *Clostridium pasteuranium* work in acid soil. (*Anaerobic* means that it does not depend on oxygen for survival).

The symbiotic bacteria are the most important nitrogen-fixing organisms. Several kinds of bacteria called *Rhizobium* have a cooperative arrangement with a group of plants called legumes, such as peas, beans, and alfalfa, whereby the two dissimilar kinds of organisms jointly manufacture nitrate from nitrogen gas. To accomplish this, *Rhizobium* invades the roots of the pea plant or other legume, where it causes innumerable swellings or nodules packed with bacteria. Without *Rhizobium*, the root tissue of the pea plant could not produce nitrate.

Man has devised a way for nitrogen fixation that is probably as important, at least in agriculture, as the sum total of electrical discharges in the atmosphere and biological activity. The most widely used industrial process for nitrogen fixation use only nitrogen and hydrogen as raw materials. Large amounts of energy are needed to compress the gas mixture to 5,000 pounds per square inch. The crucial step produces ammonia:

$$\underset{\text{nitrogen}}{N_2} + \underset{\text{hydrogen}}{3H_2} \rightarrow \underset{\text{ammonia}}{2NH_3}$$

The ammonia is used in anhydrous form as a commercial fertilizer and as a starting point for the manufacture of other nitrogen-containing products for industry and agriculture: aqua ammonia, ammonium salts, urea, and nitrates. Urea is used as a fertilizer, a cattle feed additive, and an industrial chemical.

Carbon Dioxide

Carbon dioxide gas is a minor constituent of the atmosphere, comprising about 0.03 per cent by volume of the air. By comparison, the air contains 667 times as much oxygen and 2,487 times as much nitrogen. Yet, without carbon dioxide no life could exist, for it is vital to the production of sugars and starches in plant by the process called *photosynthesis* and is a source of carbon, the basic building block for organic compounds in metabolic synthesis. Without photosynthesis there would be no plants, and without plants there would be no animals. Even carnivorous (flesh-eating) animals and parasites must rely for food either on herbivorous animals, which feed on green plants, or on other organisms which derive their carbohydrates from photosynthesis.

Photosynthesis

The sugars and their more complex relatives, the starches, collectively called *carbohydrates*, are produced in green plants from carbon dioxide and water. The reaction is driven by the sun which pumps energy into space in the form of electromagnetic radiation. Packets of light energy called *photons*, penetrate to the earth through the "optical window" of the atmosphere. Energy equal to 100 million Hiroshima bombs reaches the earth every day. Part of the radiation is absorbed by green plants, which use it as the energy for the reaction of photosynthesis.

Photosynthesis takes place in specialized pigments of the plant. The green pigments, in which the reaction mainly takes place, are collectively called *chlorophyll*, usually packaged in granules called *chloroplasts*, which can be seen easily under the microscope and are visible in the green colouration of plants. It is a curious fact that the chlorophylls are a type of structure called *porphyrins*, related in structure to the porphyrin in haemoglobin. Instead of iron as in haemoglobin, the porphyrin structure of the chlorophylls contains magnesium.

Light in the green part of the spectrum can be used only slightly by chlorophyll, for most of it is reflected and comprises a large part of the "colour" of foliage. The light rays that energize chlorophyll most effectively are in the blue, orange, and red parts of the visible spectrum.

The capture of solar radiation by plants and its utilization to energize the synthesis of sugar from carbon dioxide and water by photosynthesis is the most important biological exploit of nature. The biological mechanism of the process has fascinated scientists since the beginnings of plant studies. Yet the process remained one of the most puzzling problems in biological science until a series of brilliant laboratory experiments, made possible by the new techniques of radioactive tracers, revealed the biochemistry of photosynthesis.

In 1949 Melvin Calvin, a biochemist at the University of California at Berkeley, investigated the processes by which the green plant manufactures carbohydrates. Calvin and his co-workers used the radioactive isotope of carbon, carbon-14 (usually written ^{14}C), which they incorporated in molecules of carbon dioxide. Their experimental plants were placed in an atmosphere in which part of the carbon dioxide contained carbon-14. Calvin found that the reactions of photosynthesis occurred in split seconds. By stopping the process after exposure of the plants to the radioactive environment for seconds at a time, Calvin was able to follow the very first steps of photosynthesis. For this work Calvin received the 1961 Nobel Prize in chemistry. Subsequent work with both radioactive carbon and a heavy isotope of oxygen, oxygen-18 (usually written ^{18}O), has further elucidated the complicated steps in photosynthesis. The following overall equation summarizes the basic process:

$$CO_2 + H_2O \xrightarrow{+\text{Light}} C_6H_{12}O_6 + O_2$$

carbon dioxide — water — carbohydrate — oxygen

The oxygen that goes into the makeup of the carbohydrate comes from the carbon dioxide, while the oxygen that is liberated as free gas comes from the 'splitting' of water. A more complete overall equation is frequently given, although even this is a gross simplification of an intricate biochemical mechanism:

$$6CO_2 + 12H_2O \xrightarrow{+\text{Light}} C_6H_{12}O_6 + 6O_2 + 6H_2O$$

carbon dioxide — water — glucose — oxygen — water

The equation represents the conversion of electromagnetic energy (in the form of light rays) to chemical energy (stored in the form of a sugar). Glucose (dextrose), the simplest sugar, contains 670 calories of available energy per 180 grams.

The process of photosynthesis cannot be duplicated artificially. However, changes in the carbon dioxide content of the air can profoundly modify the natural process. Experiments by James Riley and Carl Hodges at the University of Arizona to determine the response of plants to enrichment of the air with carbon dioxide showed that plant growth increased with increasing carbon dioxide concentration. Plant yields were greatly increased at 2,500 ppm (parts per million). The optimum concentration for acceleration of photosynthesis with the species of plants used in the study was around 5 times the normal concentration of carbon dioxide in air. But above an optimum concentration, further increases caused a reduction in growth. The effectiveness of higher carbon dioxide concentrations in bringing about an increase in growth was largely dependent on the degree of turbulence at the surface of the leaves. It is the increase in ventilation resulting from turbulence, rather than greater exposure to light, that causes the "border effect," an increase in growth often observed in plants growing along the border of a field.

Effect on Humans

Carbon dioxide is ordinarily not critical insofar as concentration in man's environment is concerned. The carbon dioxide content of the air is about 320 ppm. A person can tolerate nearly 5,000 ppm, or more than 15 times the normal concentration, without experiencing any seriously adverse effect on respiration, although prolonged exposure to such a high concentration might prove to be deleterious.

During quiet breathing, an adult person puts out 250 to 300 millilitres of carbon dioxide a minute. Exercise increases the output. We do not have to consciously and deliberately breathe harder to get rid of it, because carbon dioxide controls the breathing rate. This is brought about by a buildup in the concentration of carbon dioxide in the blood as a result of the concentration gradient across the alveolar membrane and its effect on carbon dioxide diffusion. The respiratory center in the medulla oblongata is extremely sensitive to carbon dioxide in the blood and responds quickly to slight changes in concentration. This can be demonstrated by voluntary rapid breathing to ventilate the lungs, during which the alveolar carbon dioxide is reduced. There will be no impulse to breathe until the carbon dioxide again builds up. The action of carbon dioxide as a respiratory mediator is reinforced by chemoreceptor nerve endings in the carotid artery which are sensitive to oxygen and carbon dioxide in the blood.

Carbon dioxide and water are the ultimate waste products from the oxidation of cellular fuel in living organisms. In animals, the elimination of waste carbon dioxide is one of the chief functions of the alveolar membranes through which the gas passes from solution in the blood to the air spaces of the lung sacs and is then expired to the outside. Thus, the molecule of carbon that are taken up by plants in the form of carbon dioxide from the air, and which become the carbon-backbones or organic compounds in living tissue, again return to the atmosphere when the organisms respire or decay. This is part of the *carbon cycle* (See chapter 5).

Not all the organic material is oxidized and returned to the atmosphere immediately. Much of it becomes imbedded under layers of sediment where it may form coal, petroleum, or natural gas. The carbon in these vast underground graves of long-dead plants or animals may remain entrapped there for hundreds of millions of years before it again returns to the atmosphere in the form of carbon dioxide through the intervention of man when he burns the fossil remains for fuel.

The atmosphere is the respository for vast quantities of carbon dioxide from the burning of fuels. Plants may not be able to withdraw it fast enough to cope with the carbon dioxide expelled into the atmosphere from industrial waste. The dangers of excess carbon dioxide in the atmosphere are discussed in Chapter 8.

REFERENCES

Broecker, W.S. 1970. Man's Oxygen Reserves. *Science* 168. 1537-38.

Gamow, G. 1970. *A Planet Called Earth*. Bantam Books, New York.

3

The Biosphere

The natural environment of living things is a complex, though fragile, membrane around the earth. This part of the earth is abundantly rich in the quantity and variety of animals, plants, fungi, bacteria, and other living things. All living organisms, together with the physical and chemical components of the total environment, make up the *biosphere* (often called the *ecosphere*). It involves all the interactions between the living and nonliving components that produce, support, and affect life. Thus, the biosphere is more than the abode of living things—it is a self-sufficient biological system in which the thread of life is sustained in a dynamic state of equilibrium by a flow of energy from the nearest star, the sun. The energy input from the sun is captured, used over and over again by the web of life, and eventually deposited in the earth or emitted back into space (Fig. 3.1).

The biosphere—is remarkably constant in physical and chemical characteristics. So special are these conditions that perhaps nowhere are they exactly duplicated on any satellite among all the stars in the Universe. However, there may be many that are similar. The temperatures of bodies in space range over millions of degrees, from intense coldness to the hottest star—the Sun. On the contrary, the biosphere of the planet earth is an island of tranquillity, where the temperature is always within a few degrees of that necessary to maintain

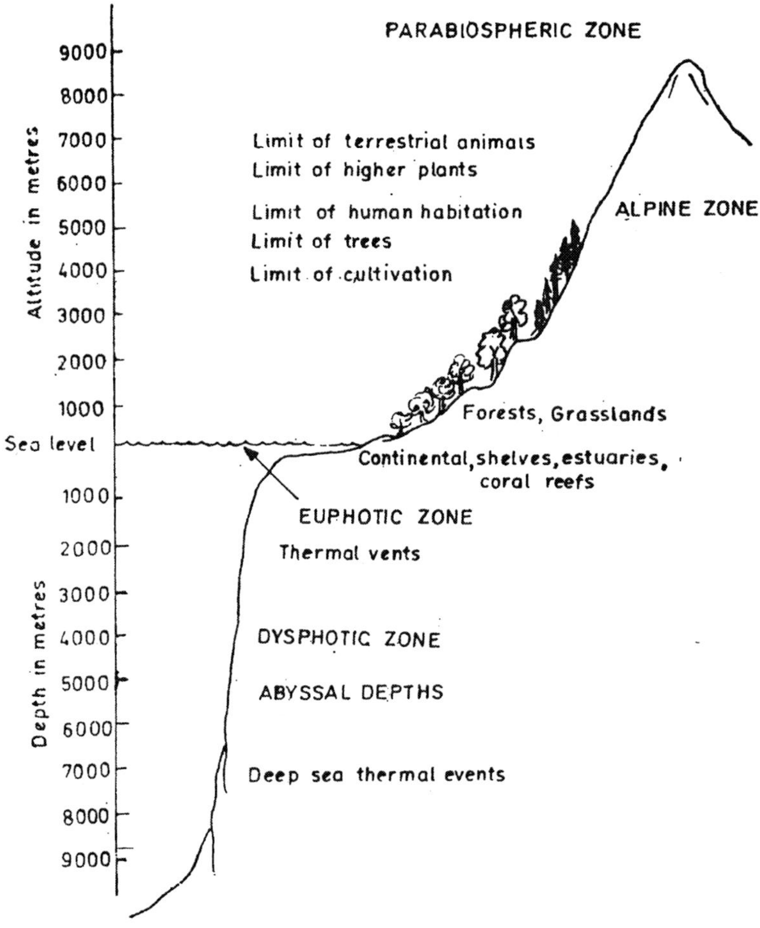

Fig. 3.1: The Vertical extent of the Biosphere

water, the giver of life, in its liquid state. The size of the earth and its gravity, rotation, and distance from the sun, as well as the chemical composition of the rocks, the ocean, and the atmosphere, hold within the membrane a delicate balance in which life can flourish. Only slightly beyond these limits, life, as we know it, would perish.

The Environment of Man

The environment of man is many things, external and internal. Physical, chemical, biological, and cultural conditions, and their ramifications, collectively comprise "the environment." Since the onset of the industrial revolution, there have been subtle but significant changes in the biosphere. These changes have become accelerated in the recent past, so that in a span of less than 100 years many forms of life have become threatened with extinction, and the health and welfare of people themselves are in jeopardy. According to reports nearly 300 species of wildlife have been listed as endangered species. This can be taken, to some degree, as an indicator of a deteriorating environment that affects man as well.

Some of the most drastic changes in the human environment are taking place as a result of the accelerating pace in the use of energy and other natural resources. The depletion of irreplaceable resources threatens to bring about drastic changes in both the world's economy and the quality of life. Yet this is only one part of the problem. Waste products are produced for which no satisfactory disposal system has been found. The solutions to these and related problems are the most difficult and challenging of human undertakings, for like other organisms, we must live within the ecological limits of the biosphere.

Man is a generalized physical animal, remarkably adaptable to changing conditions, and is highly specialized in only one attribute—a well developed central nervous system. The brain, has endowed man with a degree of cleverness unapproached by any other living thing. However, the evolutionary selection process by which man acquired cleaverness did not, simultaneously select for wisdom.

Nature seems to have played a joke on man. Without the wisdom to match his technological brains, he thrashes over and around the face of the earth, wasting, destroying, poisoning, and polluting, seemingly hypnotized by the enormity of his cleverness and oblivious to the disastrous consequences of his deeds.

Chemistry and the Biosphere

Man's activities in the material sphere have qualitatively different directions. While investigating this phenomenon as a natural event, it is convenient to distinguish two aspects of man's activities: physical (among these, thermodynamic, mechanical, electromagnetic), and chemical and biological.

For the biosphere, different aspects of man's activities are apparently, dangerous at various stages of the development of civilization. Until recent times the *biological* aspect was the most threatening: man annihilated the large animals and herded sheep and goats which literally "consumed" whole regions: he felled forests and thus brought on degradation of the top soil, promoting the advance of the desert. This factor exists even today. However, among man's disruptive activities there arose a new and now dominating aspect, the *chemical*.

Those of man's activities which directly affect the chemical form of the movement of material in the biosphere may be termed chemical activities. This aspect of man's activities is now the cause of greatest concern. For it is the chemical activity that most clearly shows up in the disruptive effects on the biosphere of 20th century civilization. It is possible that in the future this role will come to belong to the *physical* factor (primarily, to the thermodynamic, *i.e.*, disruption of the total thermal balance of the planet).

At present man's chemical activities have reached such rates that they surpass nature's capacity to neutralize them with the aid of natural feedback systems (that is, without any intervention by man). Nature can no longer wait for favours from mankind; we are obliged to help Nature.

Increase in Man-made Chemicals

Man's chemical activities which affect nature are:

— The consumption of chemical substances from nature;

— The mass introduction of substances from the depths of the earth into the environment;

— Contamination of nature by the wastes from man's activities;

— The appearance in nature and, in particular, in the biosphere of new, highly active chemical compounds separated out from natural sources or synthesized by man.

More than 6 million chemical compounds obtained form natural sources and via synthesizing procedures, are already known. Two hundred thousand substances—mainly synthetic—are added to the list each year.

Man has created a whole universe of new chemical compounds and is rapidly expanding its boundaries. Each compound has its basic physical, and chemical properties. In the universe of chemical compounds, products which are both beneficial and dangerous to the biosphere, and to man in particular, are concealed.

The data reveal that all chemical compounds possess one or another type of biological activity—that is, introduced into a living organism, they have some effect on the organism's specific properties and functions. Even seemingly inert substances such as the noble gases (helium, neon, argon, and others) have a strong narcotic effect; having entered the human organism, insoluble, inert particles of asbestos can give rise to a cancerous tumour; fine particles of silicon are capable of effectively routing some insects.

Man has created some 10,000 drugs of synthetic origin, which together with food additives generate their own type of stress, continuously and concentratedly stressing man. As a representative of a distinct biological species, what will man's reaction be to the effects of such substances not previously encountered along the paths of man's biological evolution ? It is not possible to answer this question at the present time. It is obvious that chemistry has amassed a whole arsenal of means of affecting the biosphere without, in the overwhelming majority of cases, knowledge of their powers and potentialities.

Eugene Odum (1971) has provided data which show the relationship of many respiratory ailments to contamination of the environment. In general, man's activities directed towards changing the chemical potential of the planet have led to the emergence of a whole system of uncontrolled processes, the contents and inter-connections of which are not clear in a majority of instances. Having initiated these processes, put our hopes on them, and promoted them, we condemn those consequences which turn out to be hazardous or harmful to us. However, so long as we do not understand the processes we are unable not only to regulate them but also to forecast their possible consequences.

Cause for Concern

The attainments of chemistry and industry have substantially exceeded the resources of biology and medicine to forecast scientifically the effects of chemical products on the biosphere. Mankind lacks adequate systems for tracking the chemical composition and state of the biosphere. Incomplete information about chemical changes in the biosphere on the one hand and about the biological activity of chemical compounds on the other leads to a situation in which man is not able to predict distant consequences of changes in the chemical composition of the environment.

It is only in recent times that man has begun to recognize the scale and importance of his industrial (in particular, chemical) activities. Until now man has not been ready, to the required degree, for a conscious, objectively oriented, collective effort toward optimization of this activity.

The problem posed by man's chemical activities and the need for nature preservation is vast and many-sided. Among a great number of aspects and questions are the contamination of rivers by industrial wastes, and extensive and virtually uncontrolled use of chemical fertilizers and herbicides in agriculture. There are scientific problems which are far beyond the present state of man's knowledge such as, the widespread use of antibiotics; there are problems which are obvious insofar as objectives but obscure insofar as method of implementation *e.g.*, the creation of an adequately diversified and safe complex of contraceptive agents.

REFERENCES

Hutchinson, G.E. 1970. The Biosphere. *Scientific American*, 223(3), 45-53.

Jesson, N.M. 1970. *Biosphere. A Study of Life*. Prentice Hall, Englewood Cliffs, N.J.

Odum, E.P. 1971. *Foundations of Ecology*. W.B. Saunders, Philadelphia.

Southwick, C.H. 1976. *Ecology and the Quality of Our Environment* (2nd Ed.). William Grant Press, Boston.

4

The Hydrosphere

Water is the most useful natural resource on earth, economically, culturally, and biologically. We drink it, bathe in it, relax in it, fish in it, cook in it, cool with it, irrigate plants with it, and use it for energy, power, transportation, and recreation. Though water is seemingly abundant, the uneven distribution of usable water creates a serious conservation problem in many places where it is vitally needed. In such areas the purity of the water becomes critical. Rivers and lakes in highly industrialized locations may carry an intolerable burden of chemical and human waste products to the point that aquatic life in its natural habitat is wiped out and human health is threatened. Even waters used for irrigation of crops may have excessively high concentrations of salts as the result of leaching or, along coastal areas, from underground seawater intrusion. Wells in such areas become nearly useless as sources of water for agricultural or domestic purposes.

About 300 B.C., Alexander the Great and his army of tough adventurers thrust deeply into Persia and the unknown land beyond. They came to a mighty river with more than twice the flow of the Nile, and on its banks was an ancient civilization. A thousand years before Alexander, the Aryans had invaded the continent and settled on the river. They did not give it a particular name but called it the Indus, the Aryan word for "river".

The conquered land became India, the "land of the river." The Indus was destined to irrigate 23 million acres of land, the largest irrigated agricultural region on earth. Civilizations sprang up along the other waterways. In Egypt, the middle East, India and China, as well as in South and Central America, ancient civilizations had their origins in the development of the water supply for irrigation and the production of a dependable food supply.

Modern civilization is dependent on water for irrigation, industry, domestic needs, shipping, and, of increasing importance for sanitation and waste disposal. Most of the areas of the world that are without developed water remain in the hunting and gathering or grazing stage Civilization's further advance, and possibly the survival of our cultures, will require intensive study of the development of water supplies and careful attention to the protection of water quantity and quality on a worldwide scale.

The Wet Planet

A man in space can look down on the earth and see that the surface of the planet is mainly an ocean. It is continuous except for interruptions by numerous islands. If Mount Everest, standing at 29,028 feet, were put into the deepest spot in the ocean, its peak would be more than a mile beneath the surface. The ocean occupies 70 per cent of the surface and contains 97 per cent of all the water on earth. Much of the remainder is frozen in the icecaps and glaciers. By comparison, the water in rivers and lakes is small. Less than 1 per cent is in the form of ice-free fresh water in rivers, lakes, and aquifers. Yet this relatively negligible portion of the planet's water is crucially important to all forms of terrestrial and aquatic life. There is also a large underground supply of water. Much of it remains locked deep underground for long periods of time. But the soils near the surface also serve as reservoirs for enormous quantities of water, eventually lost through evaporation or by seepage into underground storage.

The Vapour State of Water

Some of the earth's water is in the atmosphere. The warmth of the sun and the air currents evaporate about 0.03 per cent of the water on the surface each year. The vapour condenses and returns as rain and snow and eventually finds its way to the bodies of water to complete the cycle (Figs. 4.1 and 4.2).

The Solid State of Water

Much of the earth's water is in cold storage. Glaciers and the icecaps cover 11 per cent of the world's land area; icebergs and pack ice occupy 25 per cent of the ocean area. Permafrost—permanently frozen ground—holds another 10 per cent of the land area in its grip, while 30 to 50 per cent of the land is covered with snow at any given time. Three-fourths of all fresh water is locked up as ice, mostly in Antarctica and Greenland. The cold regions of the earth contain vast resources in minerals, petroleum, timber, and water. For this reason the rapidly increasing demand will surely bring more intensive development and industrial activity in those areas. However, environmental problems of the cold regions differ greatly from those in the temperate and tropical parts of the world. Disturbances of the environment that may be mildly disruptive in tropical and temperate areas could be excessively damaging in cold regions.

Small changes in world climate drastically affect the earth's distribution of water supplies. Glaciers are highly sensitive to climatic trends. Observations indicate that many alpine and valley glaciers have been shrinking during the past 100 years, although some have been increasing. As yet, there is not enough data to prove whether the major ice sheets are shrinking, growing, or in a state of equilibrium. The information might provide a means for predicting future trends. Drilling ice cores in glaciers to compare the water, dust, and isotope

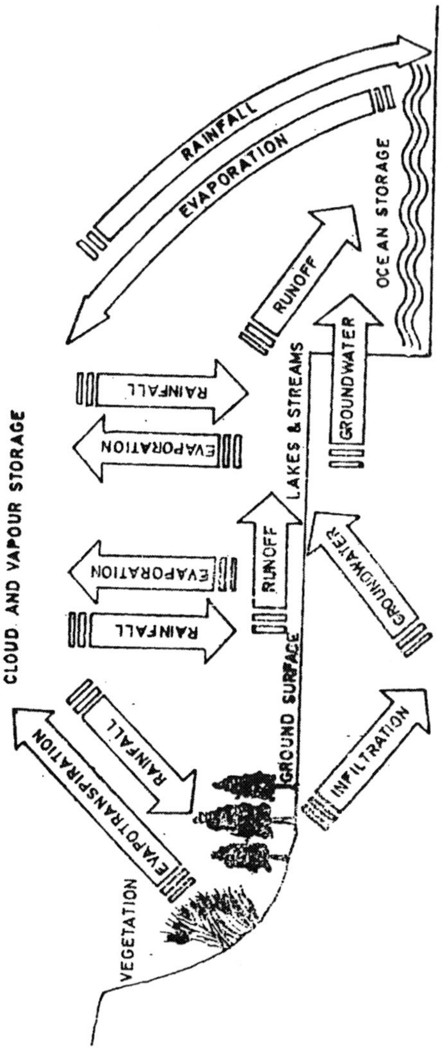

Fig. 4.1. The hydrological cycle

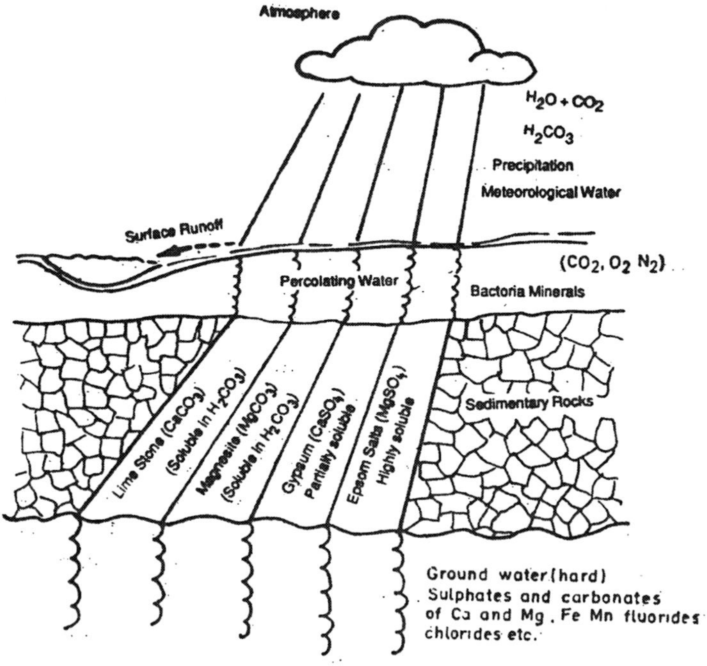

Fig. 4.2 : Relationships between surface waters and ground waters.

content of different layers makes it possible to determine precipitation during past centuries. For further discussion of climate see Chapter 9.

Liquid of Life

Water is the essence of the living process. It is the most abundant and the most versatile of the chemicals of life. It is not surprising that in its physical and chemical properties alone, water is the most uniquely remarkable substance known. Its importance to life demands an understanding of its role in the process of living organisms and in human life.

Water forms most of our flesh and blood. Protoplasm is mostly water. The content varies in different tissues of the organism and in different plants and animals. In jellyfish, the protoplasm contains 95 per cent water. Human blood is 90 per cent water. Muscles and nerves contain 80 per cent or more water. The bones contain smaller amounts of water, and even the enamel of the teeth contains 2 per cent water. The total human body consists of 60 per cent to 90 per cent water. The variability being due to the fact that the tissues tend to become dehydrated with age.

With rare exceptions, animals and plants must have an abundant supply of moisture. Animals and plants that are fugitives from the sea and have managed to escape from the limitations of an aqueous existence cannot survive for long without access to moisture either in the air, the soil, or in food. Some organisms die within seconds if deprived of water. Others must have it within their systems but can survive for years in a dessicated environment. Some animals, for example certain rodents, can manufacture their own water. Humans can with stand a loss of no more than 20 per cent of the water in the body without suffering an agonizing death. A healthy person can normally withstand no more than 7 to 10 days without water, but survival time depends largely on the amount of exertion and loss of water from heat and exposure.

The "Universal" Solvent

It is common knowledge that water easily dissolves sugar and salt as well as many other substances. Lakes, streams, rivers, springs, and seas are not pure water but contain a great deal of dissolved substances in solution. The ocean contains an average of about 3.5 per cent dissolved solids derived from the land through billions of years of erosion and run-off. The material in solution is mostly common salt (sodium chloride), but there is also an abundance of other mineral constituents, such as calcium, magnesium, sulphates, carbonates, and many metallic elements. Their concentrations are crucially important to the forms of plant and animal life that either select the water as their habitat or merely consume it for survival. The ability to take up and carry materials that are normally solid substances is perhaps the most important property of water in its role as the solvent. Water is the best solvent known. The solvent power of water enables it to carry the vital chemicals of life—minerals, salts, amino acids, and other organic substances—to the cells of the body. The return trip may be made with a load of waste products, and the water may be eliminated along with the refuse.

Terrestrial plants cannot take up mineral nutrients from the soil in the absence of water. Phytoplankton cannot absorb their mineral requirements unless the minerals are dissolved in the water of their environs. Food must be dissolved before it can enter the blood stream of animals. Even the oxygen and carbon dioxide needed by aquatic and marine organisms must be made available to most of them in solution. The products of metabolism are transported within plants in the aqueous solution called sap and within animals they are dissolved in the water of plasma. Many of the waste products of metabolism are carried away dissolved in water.

Transparency of Water

In the oceans, as on land, photosynthesis is the first stage in the nourishment of the food chain. Phytoplankton,* like land vegetation, thrive on sunlight, carbon dioxide, and mineral nutrients. Fortunately, water is transparent to light. Yet photosynthesis by the producers in the ocean is limited by the relatively poor penetration of light through water compared to air. Some light penetrates clean ocean water to a depth of about 60 metres. However, most of the photosynthesis takes place in the upper few feet. In clear ocean water, the blue and

***Phyton* (Gr)="plant"; *plangktos* (Gr)="wandering." Hence, phytoplankton are marine or fresh-water plants with weak locomotary power that drift with the surrounding water.

green portions of sunlight penetrate the best, but these are the wavelengths of visible light that are the least efficient in photosynthesis. Red light is selectively absorbed and therefore is effective only in a narrow zone very near the surface.

Clear water, such as that found in much of the open seas and in many tropical waters, is relatively free of plankton, but there is the compensating factor that light penetration is also greater, resulting in photosynthesis taking place at greater depths. Tropical waters have a high rate of biological turnover because of the higher temperatures. But production of plankton is generaly greater in cool waters than in tropical waters. A positive factor is the relatively high solubility of carbon dioxide and oxygen in water of low temperatures.

Ice

It is well known that ice floats on water. This is strange behaviour for a chemical substance but it is of great biological importance. Most materials that can be frozen and melted are heavier in their cold or solid state than in their warmer, liquid state. For example, a chunk of iron will sink in a pot of the molten metal.

The ability of ice to float is explained by the molecular behaviour of water. The molecules of water are constantly vibrating. When the water is cooled, the vibrations gradually slow down. As cooling proceeds, the molecules crowd together and form an increasingly dense pack. When a temperature of 4°C is reached, water is in its most dense condition. Below this point a sudden change takes place. The depressed molecular vibration combined with the strengthened attraction of the *hydrogen bond* causes the molecules to shift their positions to a geometric formation that forms a light, expansive latticework structure. As the water cools further, it continues to expand. When it freezes into ice, it is less dense than water in its liquid form and readily floats instead of sinking to the bottom. This can be explained as follows. As water is cooled, it becomes denser because the movement of the water molecules slow down as temperature drops. Water reaches its maximum density at 4°C. But as the temperature drops further from 4° to 0°C, the attraction between the electropositive hydrogen atoms and the electronegative oxygen atoms—called hydrogen bonding—increases, and the water molecules become bonded together in a crystalline lattice, forming ice. Because of the greater space between the molecules, ice is less dense than water and occupies about 9 per cent more volume.

The fact that frozen water floats instead of sinking to the bottom is a profoundly favourable factor for life on the earth. If ice were heavier than liquid water, it would sink to the bottom of rivers, lakes, and frozen seas where it would not receive enough heat from the sun to melt it. Much of the water on earth would be solid ice. Large quantities of the earth's water would be entrapped in an unusable form. Evaporation and precipitation would be greatly reduced. Without moisture in the air, there would be little moderating effect on the sun's radiation and there would be extreme fluctuations in temperature. The world's climate would be drastically altered. Life would be difficult if not impossible. The biosphere as we know it would not exist.

Heating and Cooling

When we heat water, its temperature increases until it boils. Continued heating causes no further increase in temperature. The average velocity of the molecules remains the same no matter how hot the vessel. All additional heat is absorbed, and the energy is put to work in breaking up the hydrogen bonds between the molecules, which must take place before the molecules can evaporate.

When the vapour returns to the liquid state it must give up its heat of evaporation. If 1 gram of steam condenses at 100°C, it gives off 540 calories (2.2 kilo Joules) of heat, exactly what it absorbed when it evaporated. This principle accounts for the warmth imparted to objects when vapour condenses on them, and the cooling effect (withdrawal of heat) when moisture evaporates from a surface. This has an important moderating effect in biological systems such as the functioning of the sweat glands; when the temperature rises, the cooling effect of evaporation prevents an excessive increase in body temperature. Evaporation is also important for plants, and accounts in part for the cooling effect in areas of vegetation. It has a cooling effect on moist soil and prevents the surface from becoming as hot as it otherwise would from the direct rays of the sun. It has a similar homeostatic influence on the temperature of bodies of water and thus has an important moderating effect on land temperatures and world climate.

Specific Heat

The ability of water to store heat is a characteristic that accounts for much of its biological importance. Water can absorb great amounts of heat while the temperature increase is very little. If we walk bare feet over sand, rock, or pavement on a hot day, the heat may soon become unbearable and we will be relieved to step into a pool of water. Although the water has received the same amount of the sun's radiation, it remains refreshingly cool. On the other hand, in the evening the sand and pavement lose their heat but the water stays about the same temperature that it was during the heat of the day. Upon cooling, the temperature of sand drops five times faster than that of water. This great capacity of water to absorb heat, the slowness of water to warm up and cool off, and its ability to give up great quantities of stored heat is summed up in a property called its *specific heat*. The specific heat of a substance is the number of calories required to raise the temperature of 1 gram of the substance through 1°C.

The water in a lake or in the sea gives up 5 times as much heat as the same amount of soil, sand, or rock. That is why a large body of water has a moderating effect on the temperature of the surrounding area. During hot weather when the water receives large amounts of the sun's radiation, the water absorbs great quantities of heat. Equally large quantities of heat are given back to the air during cold weather. Much of the life on earth is dependent on the moderating effect produced by the three-fourths of the globe that is covered by water.

REFERENCES

Kvenen, P. H. 1963. *Realms of Water.* John Wiley & Sons. New York.

Press, F. and R. Siever. 1974. *Earth.* W. H. Freeman and Company, San Framcisco.

5

The Biogeosphere

The Biogeophysical Environment

The nature of the biosphere and its relationship to living organisms are determined in large measure by the physical characteristics of the earth. The earth is the fifth largest planet in the solar system, having a diameter of slightly less than 12,800 km. Its mass would be 5.52 times as great as an equal volume of water. This is called its *density*. The surface of the earth is slightly irregular, although this would be hard to discern by an observer in space.

Lithosphere

The land part of the earth is called the *lithosphere*, from the Greek would *lithos*, meaninig "stone." The solid part, literally "stonesphere," extends 30 or 50 km beneath the surface. Beneath that, a zone called the *mantle* consists of heavy rock-like material that flows like sticky gum when under pressure. The deeper the rock, the hotter it is. The core at the center is either molten or, as believed by some geophysicists, so highly compressed that it consists of solid metal, mainly iron and nickel.

The outer portion of the earth is sometimes called the *crust* It consists of two layers, the *subcontinental* or inner layer and the *continental* or outer layer. The subcontinental layer lies beneath the continents and makes up the ocean floor. It consists mainly of a type of rock called basalt. The continental layer forms all the continents. It is mostly a type of rock called granite.

About 70 per cent of the earth's surface is water and about 30 per cent is land. Slightly less than 10 per cent of the land area is in Antarctica, too cold to be populated by man. The soils of the crust were made from erosion of the rocks by the action of water, ice, and wind. Glaciers tear away pieces of rock, winds literally sand-blast the surfaces, and rivers wear down rock and soil to deposit millions of tons of rocks, mud, sand, and silt in the oceans and other bodies of water every day.

The earth is constantly changing, not only by the erosion of mountains but also by internal distortions that lift parts of the crust to higher levels. Thus, while the material of the mountains is slowly ground away and deposited in the oceans, some of the same material again becomes part of the dry land of mountains and plains.

The history of the earth is written in its crust. Twisting, tilting, upheaval, and distortion clearly show the release of tensions in the earth that changed the surface drastically over millions of years. Volcanic eruptions repeatedly laid new rock over the old. The rocks show the record of glaciers that advanced and receded several times and covered large parts of North America and Europe. There is strong evidence that entire continents, once part of the same land mass, drifted apart over a long period of time. This is called *continental drift.* The history of life on earth from the emergence of primitive plants and animals to the appearance of advanced forms contemporary with man is imbedded in the rocks that were formed by sediments containing the remains of living forms.

The Biogeochemical Environment

The earth's crust contains all of the 88 natural elements in various combinations. The most abundant elements are oxygen and silicon. They comprise 46 per cent and 28 per cent, respectively, of the total composition of the crust. The other elements are present in smaller amounts (Table 5.1). Analysis of the average granite rock shows that it consists of 67 per cent silicon dioxide (SiO_2), 15 per cent aluminum oxide, and about 14 per cent oxides of iron, calcium, sodium, and potassium. The remaining elements make up less than 4 per cent of the rock.

TABLE 5.1. **Elements in the Earth's Crust**

Elements	*Uses*	*Abundance in Earth's Crust parts per million (ppm)*
Aluminium		81,300
Iron		50,000
Magnesium		20,900
Chromium	Stainless steel	200
Nickel	Stainless steel	80
Tungsten	Abrasion resistance	69
Copper	Electrical wire	70
Lead	Storage batteries	16
Zinc	Galvanized steel	220
Tin	Tinplate for cans	40
Molybdenum	Hardening steel	15
Mercury	Chlorine production	0.5
Silver	Photographic film	0.1

The soils of the earth are created by the decomposition of rocks. They furnish much of the nourishment for all life on land. Most of the green plants are dependent on the soil for the primary nutrients: nitrogen, phosphorus, and potassium. The abundance of these minerals in any particular place determines in large measure the productivity of the soil. A critical factor in the growth of plants is the availability of the so-called micronutrients, such as zinc, iron, and manganese. When even one of these essential elements is present in insufficient amounts or in a form that is not readily available, plants do not thrive. When deficiencies exist, all the living components of the ecosystem are affected.

Organisms are not only dependent on the geochemical environment; they modify it in many ways. For example, plants growing in a squarefoot plot of soil build it into a microcosm entirely different from its original state. Small plants grow in the shelter of the larger ones; fungi and insects proliferate in the discarded plant parts; worms find a congenial home in the organically impregnated soil; and bacteria and other micro-organisms multiply. The soil changes in structure and in nutrient content. The decomposed rock becomes a biological metropolis teeming with visible and invisible life. Similar changes take place in forests and valleys but on a grander scale. Conversely, the removal of plant growth by whatever means produces drastic changes. The soil becomes compacted and hard. Its capacity to absorb moisture is reduced, increasing run-off and erosion. Nutrients are lost, some of them irretrievably. Plants can be reestablished only with great difficulty. Having no plants, the land is shunned by animals and micro-organisms that depend on plant life for sustenance.

Trace Elements

Many of elements are present in rocks in small quantities and are, therefore, usually components of soils and water in only trace amounts. For this reason, they are often called *trace elements*, even though they may occasionally be found concentrated in large deposits. A number of the trace elements are essential to the growth, development, and health of plants. Some of them are needed in such minute amounts that they are called micronutrients.

Several of the trace elements are also essential for the growth and development of animals. Iron, manganese, and iodine are needed in small amounts. Others are needed in even smaller quantities. These include copper, zinc, molybdenum, and cobalt. Elements such as vanadium, barium, strontium, silicon, chromium and nickel are required by some species of animals. Still other elements, although present in plant and animal tissues, do not have any known biological function. Such elements are lead, mercury, cadmium, aluminum, and tin.

Some of the elements that are essential for the growth and development of plants and animals can be highly injurious when they are present in the environment at excessively high concentrations, and there is often a narrow margin between the beneficial and injurious amounts.

The Carbon Cycle

The main sources of carbon dioxide to the atmosphere are the decomposition of organic matter by micro-organisms, gas exchange in the ocean, deforestation, respiration by animals, and the burning of coal and petroleum (Fig. 5.1). Photosynthesis converts CO_2 into the structural material for plants. When plants shed their leaves, the structural material is decomposed by microorganisms and is then either released as CO_2 or incorporated into soil organic matter. The organic matter may remain in the soil or it may be leached into rivers and eventually carried to the sea.

The annual release of carbon dioxide to the atmosphere from the burning of coal and petroleum is equal to about 10 percent of the carbon dioxide used during natural photosynthesis by terrestrial plants. This release has increased the amount of carbon dioxide in the global atmosphere from early nineteenth century values of approximately 265 ± 30 parts per million (ppm)by volume to 352 ppm in 1989. Estimates of the future combustion of fossil fuels based on an energy growth of between 2 and 3 percent each year imply that the atmospheric CO_2 concentration will be double sometime in the next century.

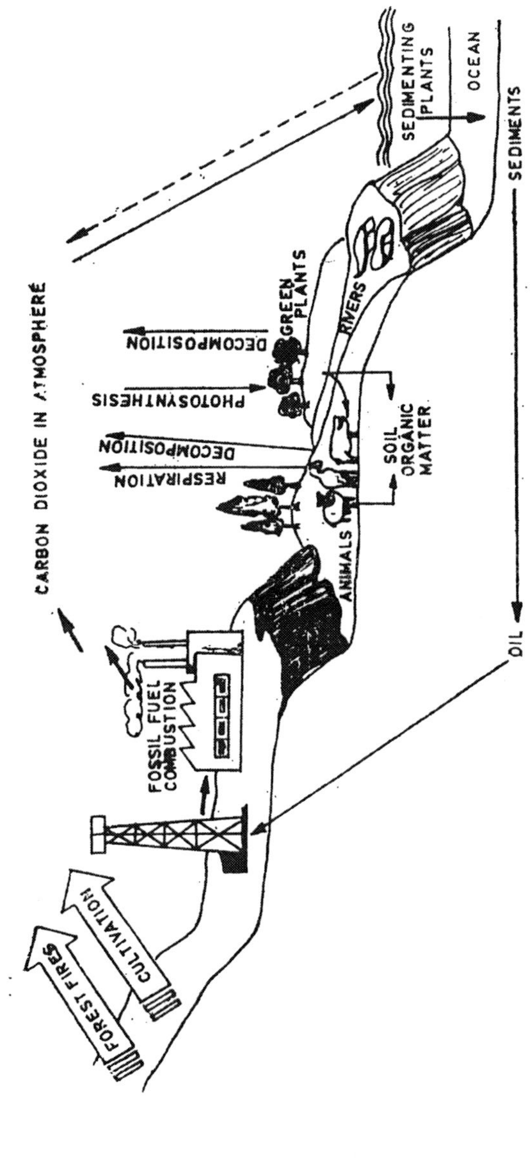

Fig. 5.1. The Carbon Cycle

Because carbon dioxide is one of the atmospheric gases important in determining the Earth's climate, a doubling of CO_2, might cause a global surface air warming. These increases are predicted to vary considerably over latitude and over the seasons. The warming of the polar regions could be two to three times as great as that in the equatorial regions, and the warming of solar winters will probably be greater than of polar summers.

In addition to CO_2, the atmospheric concentration of several other "greenhouse" gases produced by human activity also appears to be increasing. These gases are nitrous oxide (N_2O), methane, and ozone. If increases of all these gases are forecast, their synergistic effect will probably cause climate changes to occur significantly earlier than if CO_2 were acting alone. This has been discussed in detail in chapter 8.

The Sulphur Cycle

Atmospheric sulphur is derived from a number of sources, including volcanoes, the combustion of fossil fuels, and micro-organism activity in tidal flats and the water-logged soils of swamps and bogs (Fig. 5.2). The sulphur cycle perhaps best exemplifies the difficulties that impede our understanding of how man's waste emissions disturb the natural biogeochemical cycles. Sulphur is emitted into the atmosphere from natural and anthropogenic (manmade) sources, but many of the effects of both types of emissions are identical.

Anthropogenic sulphur emissions—primarily from the burning of coal and petroleum—probably exceed those from natural sources, especially in industrial regions. In the atmosphere, sulphur oxides are converted to sulphuric acid and conveyed to the Earth's surface by precipitation or dry deposition. The rapid removal of this element from the atmosphere restricts the transport of anthropogenic emissions to short distances. Unlike fossil-fuel-produced CO_2, which has a global distribution, sulphur originating from coal and

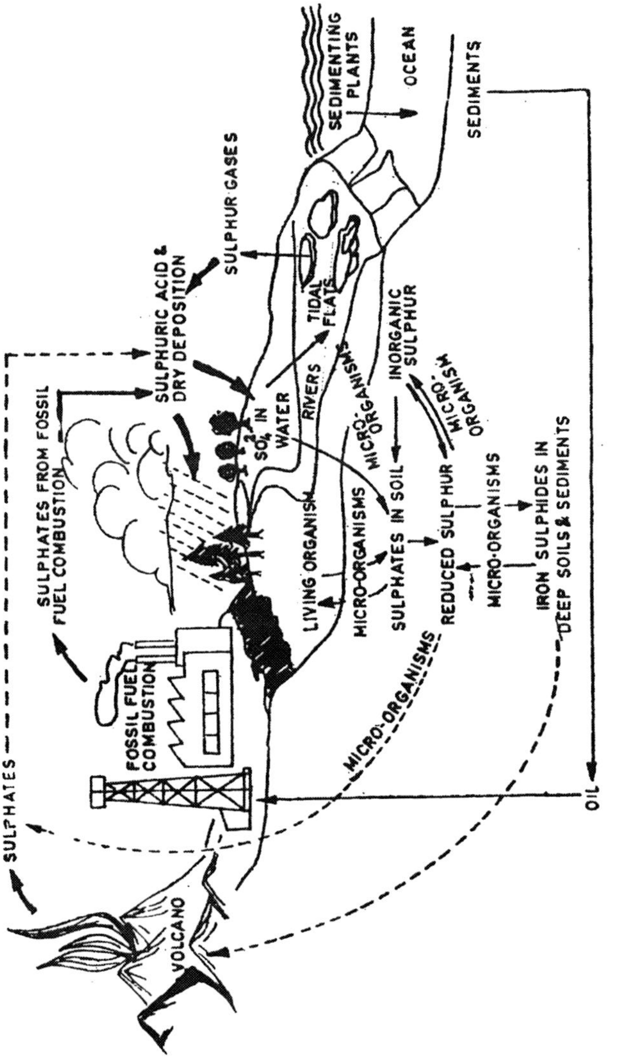

Fig. 5.2. The Sulphur Cycle.

petroleum only disperses to a maximum of a few thousand kilometres.

Sulphur dioxide and other gases (*e.g.*, ozone) have been shown to reduce plant growth near major sites of fossil fuel combustion (Johnson, 1983). This has been dealt with in detail in chapter 8 and 9).

The Nitrogen Cycle

Like carbon, nitrogen is an essential constituent of living biomass, primarily as proteins and amino acids, and changes in carbon cycling are usually accompanied by changes in nitrogen cycling. Nitrogen compounds in the atmosphere originate from natural biological activity (denitrification by bacteria, for example), volcanoes, lightning, combustion of fossil fuels, and waste products of domestic animals, as shown in Fig. 5.3.

The amounts of both nitrogen oxides and ammonia in the atmosphere are significantly influenced by combustion of fossil fuels and by wastes from domestic animal feed lots, respectively. Like the sulphur oxides, nitrogen oxides react in the atmosphere to form acidic compounds that contribute to acidic deposition. Nitrogen is also removed relatively quickly from the atmosphere. This results in impacts in industrial areas only.

The amount of nitrogen (as nitrogen oxides and nitrate) formed by the burning of coal and petroleum and also by fertilizer manufacturing is equal to about one-half of that produced naturally by the biosphere.

The Phosphorus Cycle

Unlike carbon, nitrogen, and sulphur, phosphorus does not have a significant gaseous phase in the atmosphere (Fig. 5.4). Atmospheric phosphorus is, however, largely in the form of an aerosol (fine suspended particle) that originates from crustal

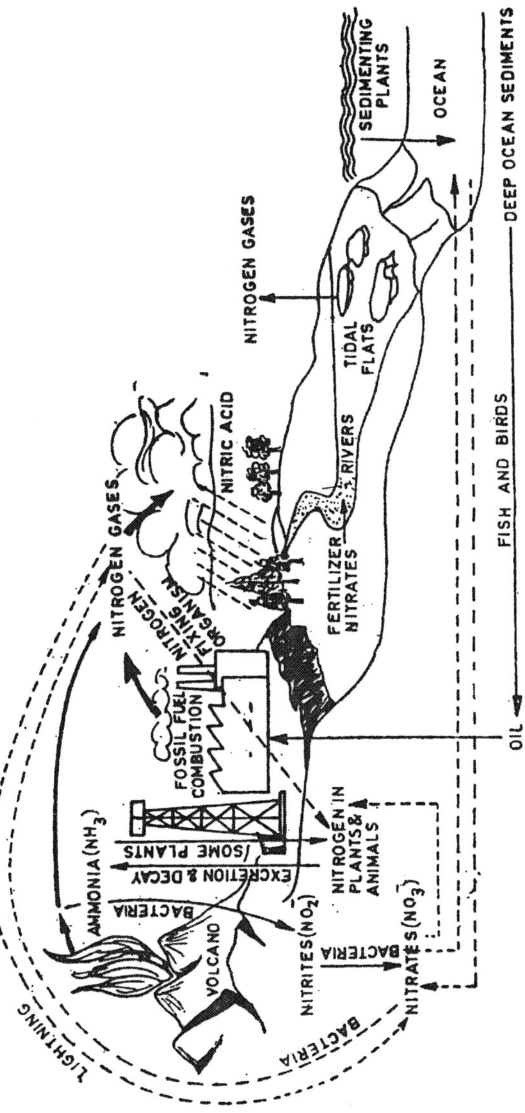

Fig. 5.3. The Nitrogen Cycle.

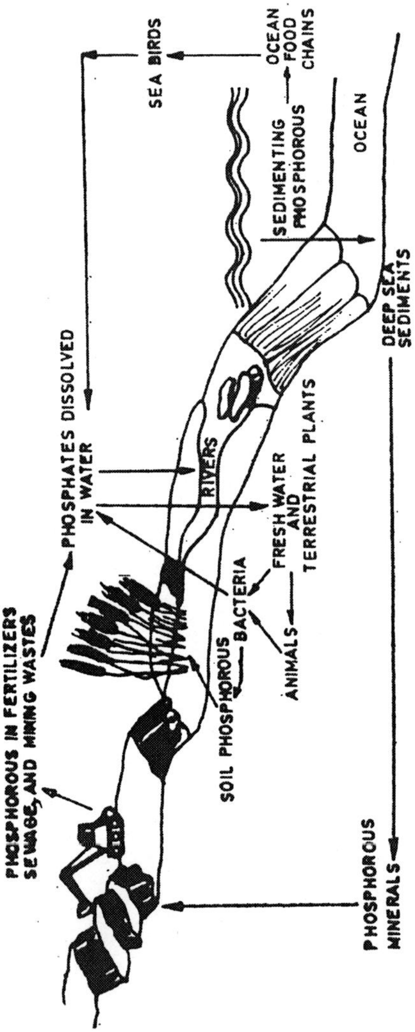

Fig. 5.4. The Phosphorus Cycle.

weathering, the oceans, and such human activities as detergent use and sewage processing.

The major impact of man on the phosphorus cycle is through the use of phosphorus fertilizers on agricultural land. Fertilizer use is necessary when the phosphorus in soil is naturally low or when it has been depleted due to intensive farming. Phosphorus is eventually removed via air or water transport to the ocean, where it is taken up by organisms and eventually incorporated into bottom sediments. After several million years, phosphorus in plant and fish remains is converted into phosphate minerals.

REFERENCES

Gamow, G., 1970. A *Planet Called Earth*. Bantam Books, New York.

Johnson, A.H. and T.G. Siccame.) 1983. 'Acid Deposition and Forest Decline'. *Environmental Science* and Technology 17, 249A-305A.

Pickering, J.S., 1961. *Captives of the Sun*. Dodd, Mead, New York.

PART II

THE QUALITY OF OUR ENVIRONMENT: CAUSES AND EFFECTS OF ENVIRONMENTAL POLLUTION

6

Air Pollution

The air is one of the most used and abused parts of the biosphere. Industrial, automotive, and domestic activities have resulted in increasingly outrageous insults to the atmosphere. Yet the air is finite; it cannot be manufactured and replenished as the need for it increases. It is at the same time our most precious and most fragile resource. Even animals and plants that live in the sea depend upon dissolved gases from the atmosphere. The enormous abuse of the atmosphere has become a health hazard and a threat to life, damaging both plants and animals in areas polluted with poisonous fumes, dust, and smoke.

Air pollution sometimes has natural causes, such as from forest, and grass fires caused by lightning. Volcanoes are a source of air pollution with ash, dust, sulphur compounds, and other gases that can be irritating or even fatal. The ash from smoking volcanoes often becomes distributed over wide areas and causes distress due to irritation of the respiratory tract and possibly aggravation of respiratory inflammations. The eruption in 1883 of the volcano on Krakatoa, an island between Sumatra and Java, blew most of the island away and started a tidal wave that killed 36,000 people. Volcanic dust from the explosion spread around the earth and was visible for many months.

Nature, however, has the power to recover from her own ravages. Following the industrial revolution, humans have caused an upsetting of nature's air ecology. If our rush for increased comforts continues as in the past, we shall not only suffocate ourselves, but also destroy all life around us. Air pollution, basically the presence of foreign substances in the air, has become a serious problem because the concentration and qualities af these substances are injurious not only to property, but also to vegetation and animal life.

Motor vehicles are by far our worst polluters, followed by industry and power plants. Carbon monoxide constitutes the greatest mass of any air pollutant, followed by sulphur dioxide, hydrocarbons, nitrogen oxides, and particulates. Most air pollutants are toxic to plant and animal life. The effects are discussed separately in subsequent chapters.

The balance of pollutants in the atmosphere is illustrated in Fig. 6.1; the direction of pollution are marked by dark arrows; white arrows indicate the process of purification.

Major Air Pollutants

Carbon Monoxide, CO

Carbon monoxide, a colourless, odourless poisonous gas, is the product of incomplete combustion of carbon and its compounds. Created largely by the automobile, CO constitutes more than half the pollution caused by humans. Carbon monoxide is very dangerous to humans because it combines with haemoglobin in the blood to form carboxyhaemoglobin. If taken in sufficient quantities, CO prevents the haemoglobin from carrying the needed oxygen to the tissues. (For details see Chapter 11.)

On an annual basis, about ten times as much carbon monoxide enters the atmosphere from natural sources as from transportation and other human activities. Most of this (77.6 percent) comes from atmospheric oxidation of methane arising from the

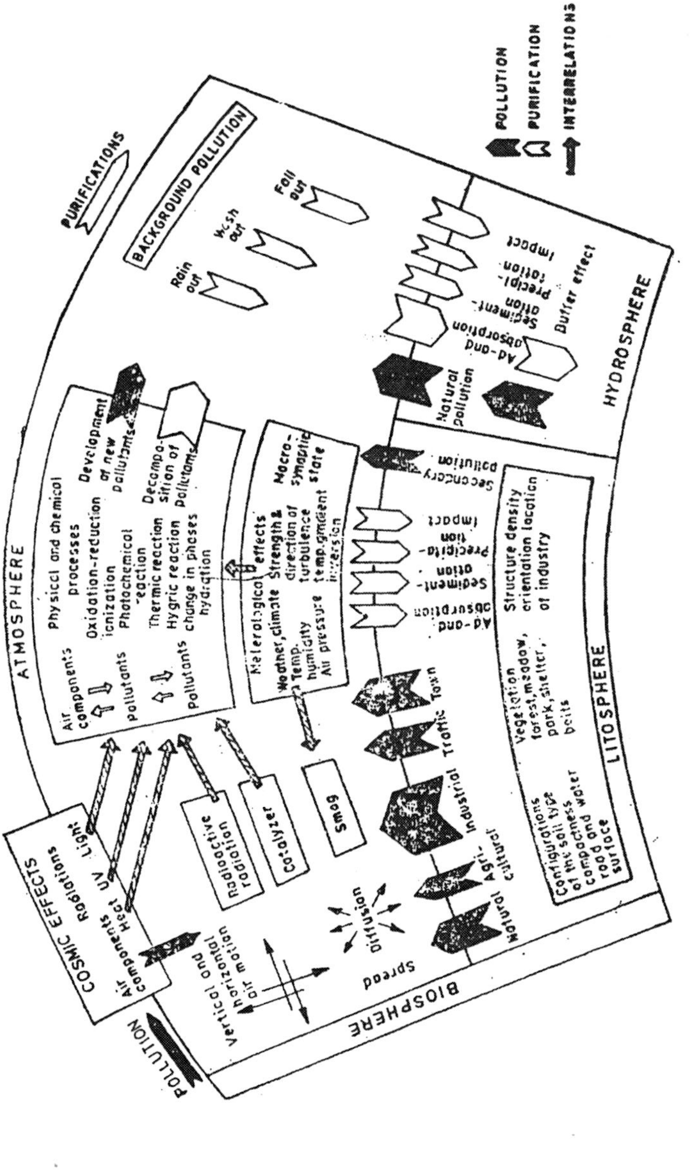

Fig. 6.1. The balance of pollutants in the atmosphere

breakdown of organic residues, especially in tropical regions. Smaller contributions come from the oceans (3.9 per cent) and from the growth and breakdown of chlorophyll in plants (2.6 per cent). This naturally generated carbon monoxide does not pose any serious problems because the rate of emission at any given place is relatively low. However, man-made carbon monoxide tends to be released rapidly in urban areas. This results in localized concentrations 50 to 100 times higher than the global average.

The concentration of carbon monoxide in the atmosphere at any time depends not only on the rate of production but also on the rate of removal. This occurs mainly in the soil, though a small amount is oxidized to carbon dioxide in the lower atmosphere. In the soil, fourteen species of fungi have been identified as the active agents oxidizing carbon monoxide to carbon dioxide. The activity of these fungi varies with the type of soil, being least active in desert soils and most active in the tropics. Cultivated soils are less satisfactory to these fungi than are those covered with natural vegetation. The total capacity of soil fungi for carbon monoxide oxidation seems to be more than adequate to deal with the carbon monoxide produced on a global basis. Unfortunately, the urban areas, where most anthropogenic carbon monoxide is released, have among the poorest soil reservoirs of these fungi which, therefore, cannot make an effective contribution to lowering the localized high concentrations in these areas. One study showed that in big cities, the air in heavy traffic contained more than 30 ppm carbon monoxide, a concentration that approaches the threshold for acute poisoning. Underground garages have been found to contain 100-200 ppm carbon monoxide in air.

Carbon Dioxide, CO_2

All organic substances used for fuel—petroleum, coal, natural gas, wood are the products directly or indirectly, of photosynthesis. Petroleum, natural gas, and coal are the fossil remains of ancient living organisms, modified by the conditions to which they were subjected under the pressure of sediment and rock deposited by prehistoric seas. Petroleum, natural gas, coal, and wood all contain an abundance of carbon that was once part of the atmosphere or the sea. In the form of carbon dioxide, it was taken up initially by plants in the process of photosynthesis. Since carbon dioxide and water are the principal products of the complete combustion of fuel, the organic carbon is returned to the atmosphere in its ancient form (see carbon cycle. Fig. 5.1).

Carbon dioxide comprises a minute fraction of the gases in the atmosphere—about 3 parts in 10,000. Yet it has a profound effect on the temperature of the atmosphere, and hence, on climate. This is discussed further in chapter 8. The gas is substantially soluble in water, so that the ocean contains about 1 part in 10,000. This is a fortunate circumstance for the photosynthetic activity of marine and aquatic phytoplankton. Because of the small amount of carbon dioxide in the biosphere, and its relatively low toxicity, it would not ordinarily be thought of as a pollutant. However, carbon dioxide does not enter into the photochemical reactions in the atmosphere and its removal is dependent on biological and chemical deposition. The equilibrium brought about by the normal uptake and output from natural sources during the present and recent past may be upset drastically by the enormous increase in the combustion of fuels during this century.

In the geologic past, there were large fluctuations in atmospheric carbon dioxide. Volcanic eruptions injected enormous quantities of carbon dioxide into the atmosphere. The total amount during several billion years is estimated to have been at least 40,000 times the amount now in the air. Most of the carbon dioxide was precipitated from the ocean and other bodies of water as salts of calcium or magnesium in the form of limestone or dolomite (magnesium-calcium carbonate). About one-fourth of the total carbon dioxide was taken up by plants and was buried in the sediments. A small fraction of this was concentrated in deposits of coal, oil, sands, and gas pockets that became the fossil fuels of our industrial age.

As long as man continued to use mainly wood for fuel, the contribution of carbon dioxide to the atmosphere was negligible and burning affected the normal carbon dioxide cycle only slightly. Now, however, man is burning fossil fuels that were deposited by living organisms during the past several hundred million years. These geologic reserves of locked up carbon fuel are being oxidized for industrial purposes at a rate that is causing measurable and significant changes in the carbon dioxide content of the atmosphere.

Global Carbon Dioxide Concentrations

About 10 times as much carbon dioxide is injected into the atmosphere from coal fires and furnaces as from the process of breathing. These sources alone would double the carbon dioxide within 500 years if there were no natural process of removal. If the recoverable reserves of fossil fuels—coal, lignite, petroleum, natural gas,—are 2,971 billion metric tons, the the carbon dioxide equivalent of the reserves is more than 300 per cent of all the carbon dioxide in the atmosphere present in 1950.

The production of carbon dioxide since 1950 has averaged 9 billion tons per year, resulting in an increase in concentration of about 1.6 ppm per year. At this rate, it is estimated that by the year 2000 there will be about 25 per cent more carbon dioxide in the atmosphere than at present. But the use of fossil fuels is increasing. Estimates of the rate of increase in combustion vary between 3.2 per cent and 5 per cent per year. If the higher rate continues without further acceleration, the quantity of carbon dioxide that would enter the atmosphere during the next 30 years will be an astounding 60 per cent of all the carbon dioxide that was in the atmosphere in 1950. And if the current rate of combustion were continued, about 13,000 billion metric tons of fossil fuel would be used up during the next 1,000 years, and would add 17 times as much carbon dioxide as the present amount. (See chapter 8 for more details)

Physical Effects of Carbon Dioxide

Building stones, especially limestone materials such as.

marble, are susceptible to acid-forming substances. The calcium carbonate of limestone is converted by carbon dioxide and water to bicarbonate, which is water soluble. "Stone cancer" is a term that has been used to describe the erosion of buildings and art objects in the polluted atmosphere of Venice, Italy. Much of the loss there, however, may also be the result of ılphur and sulphide combustion products in the industrial effluent.

Effects on Animals and Plants

Although carbon dioxide is essential to the physiology of animals, and more especially to that of plants, the concentration is normally not critical. For example, uncontaminated air has a carbon dioxide content of about 300 ppm yet man can tolerate nearly 5,000 ppm without drastically adverse effects on respiration. Moderately higher concentrations than normal bring about increases in the rate of photosynthesis in plants, but an increase of several fold is required before deleterious effects are noted. However, carbon dioxide has a physical role in the stability of the biosphere that may be as important, indirectly, as the direct effect on animal and plant physiology. Carbon dioxide forms a blanket of insulation around the earth. Changes in temperature caused by an increase in the carbon dioxide content of the atmosphere could alter world climate in ways that would drastically affect all living organisms. It is theorized that these changes can be caused by a greenhouse-like effect. (See chapter 8 for more details)

Sulphur Oxides, SO_x

Sulphur pollutants are belched into the air in enormous quantities as part of the industrial effluent. Atmospheric sulphur comes mainly from the sulphur content of fuels. Nearly 80 per cent of the sulphur in sulphur dioxide is initially emitted as hydrogen sulphide and is converted to sulphur dioxide in the atmosphere. Sulphur dioxide and the products of its reactions are highly damaging to structural and other materials, as well as being physiologically harmful and injurious to health. Sulphur pollutants were probably the predominant pollutants in most of the "killer smogs."

Sulphur dioxide is a colourless gas with a penetrating and pungent odour. It produces respiratory irritation at low concentrations. As an industrial air pollutants, it is formed by the oxidation of the sulphur compounds in fuel. The gas is moderately soluble, and in the presence of moisture it forms sulphurous acid, a fair oxidizing agent:

$$\underset{\text{sulphur dioxide}}{SO_2} + \underset{\text{water}}{H_2O} \rightarrow \underset{\text{sulphurous acid}}{H_2SO_3}$$

Sulphurous acid is converted slowly to sulphuric acid (H_2SO_4). Sulphur dioxide also reacts very showly in oxygen to form sulphur trioxide gas (SO_3), which in turn reacts quickly with water, forming sulphuric acid. Aerosols (fogs) of sulphuric acid contribute to the haze from the combustion of high sulphur fuels in industrial areas. The bluish white plume often seen coming from the stacks of electric power plants is often largely a sulphuric acid aerosol.

Sources of Sulphur Oxides

The major natural sources of sulphur dioxide are volcanic in origin. However, this contribution to pollution is believed to be small. Most of the total emission of sulphur oxides is from burning of fossil fuel by industries, and furnaces. Fossil-fuel power plants, for generating electricity, are the major source, and they are being built more rapidly than anticipated because nuclear power plants are being installed more slowly than planned. Most of the remaining sulphur oxide emissions is accounted for by nonferrous smelters and petroleum refineries. The major sources are highly varied in size and design, and reduction of their emissions to acceptable levels requires a diversity of highly complex control systems. Emissions are increasing at an alarming rate.

In India sulphur dioxide emissions are maximum (120 tons per day) in Calcutta followed by Bombay, Delhi, Ahmedabad, Kanpur, Hyderabad, Madras, Nagpur and Jaipur in decreasing order.

Physical Damage

Sulphur dioxide and its end products have a severe corrosive and deteriorating effect on many materials. Sulphur dioxide and sulphur trioxide are especially damaging to building materials that contain carbonates, such as limestone, marble, mortar, and roofing slate. The reaction forms soluble calcium salts that are washed away by condensation and rainwater. Works of art in or near industrial areas and cities are in serious jeopardy. One of the two specimens of Cleopatra's Needle was moved to New York's Central Park in 1880. The obelisk has deteriorated more since its arrival than during its previous 3,000 years in Egypt where it originally stood before the Temple of the Sun at Heliopolis. Some works of art are being moved indoors to preserve them.

The atmospheric deterioration of marble and limestone structures in Italy is referred to as "stone cancer." The art works and irreplaceable structures of Venice are decaying from the emissions of nearby industries. Nearly all of the thousand or so *palazzi* and churches contain material of marble having a high carbonate content, highly susceptible to sulphur oxide corrosion. Venice has other environmental problems including floods, humidity, salinity, industrial water pollution, changes in the populations of plants and animals in the lagoon, impaired tidal flow for removal of sewage, a rising water level, and sinkage of the foundations. These, combined with corrosive air pollution, make Venice appear to be a dying city.

Sulphur oxides even cause the destruction of extremely durable structural materials. Steel corrodes up to four times as fast in cities as in rural areas. Particulate matter accelerates the attack of sulphur oxides on metals, especially damaging when moisture is present. The oxide and the acid attack iron, steel, copper, nickel, zinc, and at high concentrations, even aluminium

Sulphur oxides attack some paint pigments and modify fresh paint films in a way that delays drying and increases the susceptibility to moisture. Sulphur pollutants even attack nylon. It makes paper brittle due to the presence of metallic impurities that catalyze the conversion of sulphur dioxide to sulphuric acid. Sulphur dioxide also attacks leather and textiles, causing slow deterioration. Synthetic fibers, but particularly cotton and wool, are susceptible. Laboratory experiments under controlled conditions show that photo-chemically produced compounds in the atmosphere cause fading of dyes in fabrics, and the addition of sulphur dioxide causes a synergistic effect, greatly increasing the damage.

Hydrogen Sulphide, H_2S

During the nights of November 23 and 24, 1950, in the small town of Poza Rica, Mexico, a poisonous gas drifted into homes and struck down people of all ages. The toxic gas, hydrogen sulphide, came from an industrial plant built to manufacture sulphur from natural gas. Hydrogen sulphide is ordinarily removed from natural gas as one of the steps in the process of recovering by-product sulphur, but in the early morning of November 24, the hydrogen sulphide was inadvertently released and caught many of the resident in bed. Hydrogen sulphide causes rapid respiratory irritation and central nervous system impairment, including unconsciousness, and death. In all, 320 persons were hospitalized and 22 died.

Hydrogen sulphide is a colourless gas, having about the same toxicity as cyanide. The gas has a penetrating odour resembling rotten eggs, and is therefore detectable by the human nose at very low concentrations, about 2 ppb (parts per billion). However, the olfactory sensory organs become fatigued quickly and can no longer respond. The sense of smell for hydrogen sulphide is lost within 2 to 15 minutes at 100 ppm or higher. This characteristic makes hydrogen sulphide extremely dangerous in cases of excessive or prolonged exposure.

Low concentrations of hydrogen sulphide cause headache, nausea, lassitude, collapse,* coma, and death. Hydrogen sulphide readily passes through the alveolar membrane of the lungs and penetrates to the blood stream. Death may come within a few seconds from even one or two inhalations. At 1,000 ppm it is almost instantaneously fatal. Death is from respiratory failure. The California standards for ambient air quality specify an "adverse" level of 0.1 ppm for 1 hour as sufficient to produce sensory irritation. The maximum allowable concentration for an 8-hour day is 20 ppm.

Natural Sources

The principal natural sources of hydrogen sulphide are decaying vegetation and animal material, especially in shallow aquatic and marine environments. Anaerobic bacteria attack sulphates in deep sediments, reducing the sulphate to sulphide. Anaerobic decomposition sometimes takes place in underground organic deposits, causing hydrogen sulphide to be present in domestic water supplies. Water from such sources has an objectionable odour but in some parts of Europe water containing hydrogen sulphide is considered to have medicinal properties. A concentration of 0.05 ppm is detectable by taste. Hydrogen sulphide is given off by sulphur springs, volcanic discharges, gas wells, coal pits, and sewers. It has been estimated that about 30 million tons of hydrogen sulphide are released annually by ocean areas and 60 to 80 million tons by land areas.

Industrial Sources

Industrial pollution is probably not often a significant factor in public health, although it can be a threat in localized areas and form exposure during industrial operations. The important industrial sources of hydrogen sulphide are those which also give rise to sulphur oxides and sulphuric acid, the users of

*Lassitude is a state of tiredness, laziness or lack of interest.

high-sulphur fuels. About 80 per cent of the sulphur compounds in fuels are converted to hydrogen sulphide which is soon oxidized in the atmosphere to other sulphur compounds. Notable emitters of hydrogen sulphide are petroleum refineries, kraft paper mills, tannery wastes, and mining operations.

Most of the hydrogen sulphide released to the atmosphere does not remain in that form for long. It is oxidized to elemental sulphur and sulphur dioxide. The latter is further oxidized to sulphur trioxide and subsequently to sulphuric acid and sulphates. An important reaction in the atmosphere is the oxidation of hydrogen sulphide with ozone:

$$\underset{\text{hydrogen sulphide}}{H_2S} + \underset{\text{ozone}}{O_3} \rightarrow \underset{\text{sulphur dioxide}}{SO_2} + \underset{\text{water}}{H_2O}$$

The reaction takes place fairly rapidly, with a half-life of atmospheric hydrogen sulphide of 2 to 28 hours, depending on location and conditions. The hydrogen sulphide does not remain as a pollutant for a long period of time, but is a transient toxicant. It can be highly injurious, however, if concentrations are high or if exposure is prolonged.

Aerosols

An aerosol is a suspension of finely divided solid particles or droplets of liquid in the air. Examples are smoke, fog, dust, and the condensation plume from steam. The term *particulate* is often used to refer to both solid and liquid aerosol particles in the air. The amount and nature of suspended particulates in the atmosphere and their chemical and physiological effects are among the least understood aspects of air pollution.

Source of Particulates

Particulates are not necessarily visible. Suspended particles of microscopic dust, chemicals, liquids, sea-salt, pollens, aeroallergens, bacteria, and viruses are abundantly present in atmosphere. Some of them are the cause of much human discomfort and diseases. Other injurious aerosol constituents include soot, lead particles from automobile exhaust, asbestos, fly ash, volcanic emissions, pesticides, sulphuric acid mists, and metallic dusts. Dozens of substances have been identified as minor particulate components of air, possibly including many toxic materials that have not yet been determined to be injurious.

A surprising amount of dust settles on the earth from outer space. Every day 8 billion solid particles penetrate the atmosphere. The average size is less than a grain of sand. There is enough dust and ash left over from the attrition of the larger bodies and the remains of the disintegrated material to leave an accumulation of about 2 million tons of debris that settle on the surface of the earth annually. It drifts down and is found everywhere from the high mountain peaks to the sediment at the bottom of the oceans.

Microscopic study of air samples taken in Austria showed that an enormous number of particles are dispersed in the air. The number varied widely from 5,000 per cubic centimeter in downtown Vienna to 800 particles per cubic centimetre in the residential area, as compared to 300 per cubic centimetre in natural aerosols.

Size of Particulates

Aerosol particulates in the atmosphere range in diameter from about 0.001 micron* to more than 100 microns (Fig. 6.2). Particles larger than 10 microns settle rapidly, although they may remain air-borne for moderately long periods under turbulent conditions. Particles of 0.1 to 1.0 micron would normally remain air-borne for extended periods, except that collisions and electrostatic attractions, as well as other forces, result in coagution and the formation of effectively larger particles. Particles

*A *micron*, often designated by the Greek letter mu (written μ) is 0.001 millimetre.

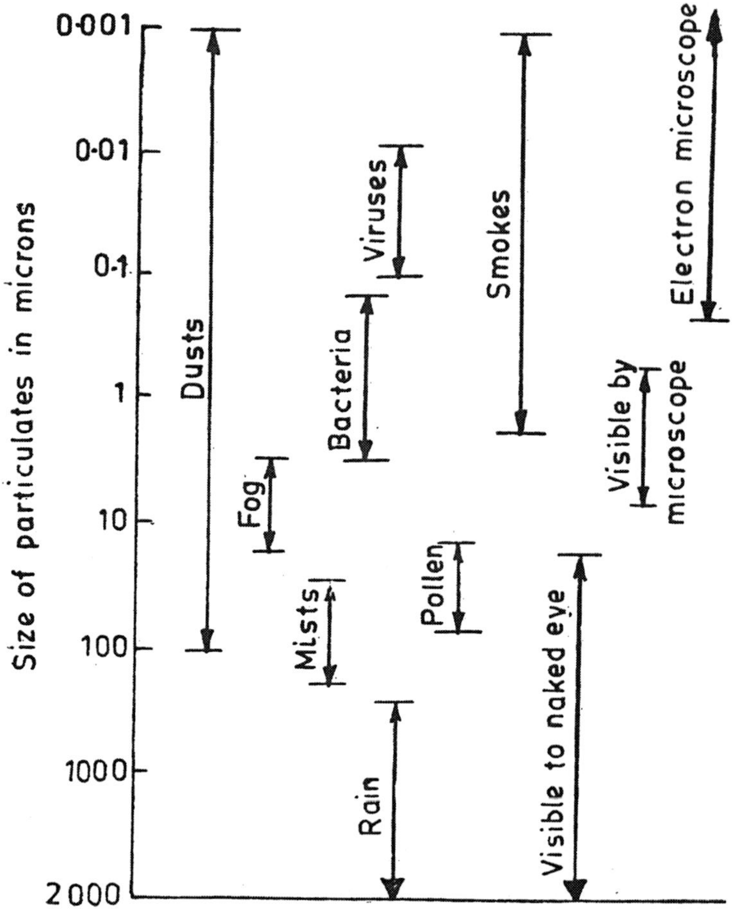

Fig. 6.2. Relative size of atmospheric partices in microns (1 micron=10^{-6} m).

of less than 0.1 micron can form extremely stable aerosols, but by random movement in the air, eventually collide with other particles and form larger particles by coagulation.

Most of the particles injected into the atmosphere fall out within minutes, hours, or days. But the smaller particles may remain air-borne for years. Many of the aerosol particles originating from human-made sources in cities and industrial areas, as well as from volcanoes, forest fires, and nuclear explosions, travel across international borders.

Some of the aerosol particles, mainly sizes above 2 microns, are removed from the air by rainfall. Particles also disappear from the air by impinging on objects such as trees and buildings. Extremely small particles do not readily impinge on objects because of their aerodynamic properties. The size judged to be most liable to enter the respiratory tract and become lodged in the air spaces of the lungs is about 1 micron.

Particles emitted from automobile exhaust are mostly less than 5 microns in diameter. They consist of carbon particles, lead compounds, and reaction products from the combustion of motor oil. Other particulates originating indirectly from automobile pollution of aerosols are formed as reaction products from nitrogen oxides and hydrocarbons in the atmosphere. Diesel engines emit more particulate material than gasoline engines. Diesel exhaust is estimated to contain 62.5 per cent particles less than 5 microns in size and 37.5 per cent in the 5 to 20-micron range.

Surface Area

The effects of small particles are best understood by considering the relationship between the size of the particles and their surface area. Assuming the shape of particles in an aerosol to be perfect spheres, the total surface area is inversely proportional to the diameter of the particles. In order to visualize this, let us take a droplet having a diameter of 100 microns and a volume of 523,600 cubic microns. Then split this large droplet into smaller droplets, each having a diameter of 10 microns. The volume will be the same, but instead of 1 droplet there will now be 1,000 droplets; and they will have a total of 10 times the surface area of the original droplet. Split these droplets further into smaller ones each having a diameter of 1 micron, and there will now be 1,000,000 droplets with a surface area 100 times the original droplet.

The enormous surface created when chemicals are divided into small droplets greatly increases the possibility that chemical reactions will take place. Large surfaces also bring about physical changes such as increased absorption of heat and light scattering.

Physical and Physiological Effects

Particulate matter has damaging physical effects, partly due to the acceleration of the corrosive effects of other pollutants such as sulphur oxides and sulphuric acid and partly due to the formation of grime. Dirt and soot are damaging to fabrics and other materials, which are then costly to clean, requiring the use of large quantities of solvents and detergents which further add to the pollution cycle.

The physiologically injurious effects of pollutants such as sulphur dioxide and its reaction products are made more severe by the presence of particulate matter. The synergistic effect was noted in the case of the London "killer smog" of 1952; the evidence indicated that it was the combination of particulates and sulphur oxides that caused 4,000 deaths during that episode.

Cancer-inducing substances have long been known to be present in particulate pollutants. They induce cancer when painted on the backs of mice or when injected under the skin but do not consistently do so when injected into the lungs. However, lung tumours, resembling human lung cancer, are produced in experimental animals when benzpyrene is mixed with iron oxide or hematite and injected into the trachea. These materials are commonly found in mixture with other atmospheric pollutants. Toxicity experiments involving injections of extracts from polluted air into newly born mice show that there is a broad spectrum of carcinogenic materials in the atmosphere.

Soots are physically complex structures, mostly highly porous agglomerates of great absorptive power. Since soots are produced in conjunction with other combustion products, many of them potent toxicants, the potential for synergistic action is inherent. Polycyclic hydrocarbons, to which group the potent carcinogen benzpyrene belongs are produced generally in combustion processes. In humans certain "inert" particulates, including those in cigarette smoke, cause resistence to the passage of air in the respiratory system.

Ozone

Ozone O_3 is a gas and is present in significant amounts in both the stratosphere and the troposphere, the combined amount being known as total column ozone. Stratospheric ozone is generated by the action of ultraviolet light on molecules of oxygen:

$$O_2 \xrightarrow{hv} O + O$$

$$O + O_2 \longrightarrow O_3$$

Tropospheric ozone is generated by the action of ultraviolet light on molecules of nitrogen dioxide:

$$NO_2 \xrightarrow{hv} NO + O$$

$$O + O_2 \longrightarrow O_3$$

Formation of tropospheric ozone is accelerated by sunlight acting on atmospheric pollutants, particularly hydrocarbon emissions from car exhausts and NO_x emissions from transport and industry. At high concentration, tropospheric ozone has deleterious effects on human health (WHO recommend an 8 h maximum in air of 60 parts per billion). Tropospheric ozone also reduces crop yields in sensitive species and damages forests. Tropospheric ozone is a potent greenhouse gas and may contribute significantly to the problem of global warming.

Ozone is an important component of so-called oxidant smog.* There is very little emission of ozone from automobile exhaust. However, ozone plays a key role in the photochemical formation of atmospheric pollutants (see Photochemical Smog). Ozone is the most reactive form of molecular oxygen and the fourth most powerful oxidizing agent known, exceeded only by fluorine (F_2), oxygen difluoride (OF_2), and atomic oxygen. Ozone has a distinctively pleasant odour at concentrations of about 2 ppm or slightly less, but it is irritating at higher concentrations. It is used as a disinfectant for air and water, and in industry for bleaching waxes and oils, as well as for organic synthesis. The use of ozone as an air disinfectant for closed spaces is dangerous.

Nitrogen Oxides, NO_x

Nitrogen combines with oxygen to form a family of oxides, collectively called nitrogen oxides (indicated by the symbol NO_x), which play a multiple role in air pollution. They enter into chemical reactions with other substances in the atmosphere with the production of additional irritating and toxic ingredients of smog.

The fixation of nitrogen is a vital biological process in certain plants and microorganisms and that industrial nitrogen fixation by the chemical processing of air is an important commercial enterprise. The automobile and other combustion-power devices are also nitrogen-fixing machines, although in combustion the immediate end product is a contaminant rather than a useful product. In petrol and diesel engines, the high

*The term *smog* is a short form of *smoke-fog*. Its composition is highly variable.

temperature and pressure in the combustion chamber provide suitable conditions for the production of nitrogen oxides, which, unfortunately, enter into a number of reactions in the atmosphere responsible for a number of air pollutants.

The nitrogen oxides are among the most toxic substances found in the atmosphere. One of the lesser toxic members, *nitrous oxide*, called "laughing gas," is usefull in medicine as a mild anaesthetic. Nitric oxide, and nitrogen dioxide, because of their abundance and toxicity, are the most dangerous in terms of public health. A third highly toxic member, *nitrogen pentoxide*, is a potential threat.

Nitric oxide, NO, a colourless gas, is the primary product of combustion of nitrogen. It is produced when the temperature is high enough to cause a reaction to take place between the nitrogen and oxygen in the air that is sucked into the cylinder with the air-fuel mixture. Thus the form of the nitrogen oxide that is emitted to the atmosphere in the exhaust is almost all nitric oxide. However, a large portion of it is converted to the more toxic nitrogen dioxide by a complex series of reactions in the atmosphere (Fig 6.3).

On contact with air, nitric oxide combines with oxygen, or even more rapidly with ozone, to form the poisonous nitrogen dioxide:

$$\underset{\text{nitric oxide}}{2NO} + \underset{\text{oxygen}}{O_2} \rightleftharpoons \underset{\text{nitrogen dioxide}}{2NO_2}$$

At high temperatures the equilibrium is reversed, converting nitrogen dioxide back to nitric oxide.

Nitrogen dioxide, NO_2 is the only widely prevalent pollutant gas that is coloured. Pure nitrogen dioxide is deep reddish brown and causes much of the atmospheric discolouration on bad smog days in metropolitan areas. Power plants and other energy conversion systems as well as a variety of chemical

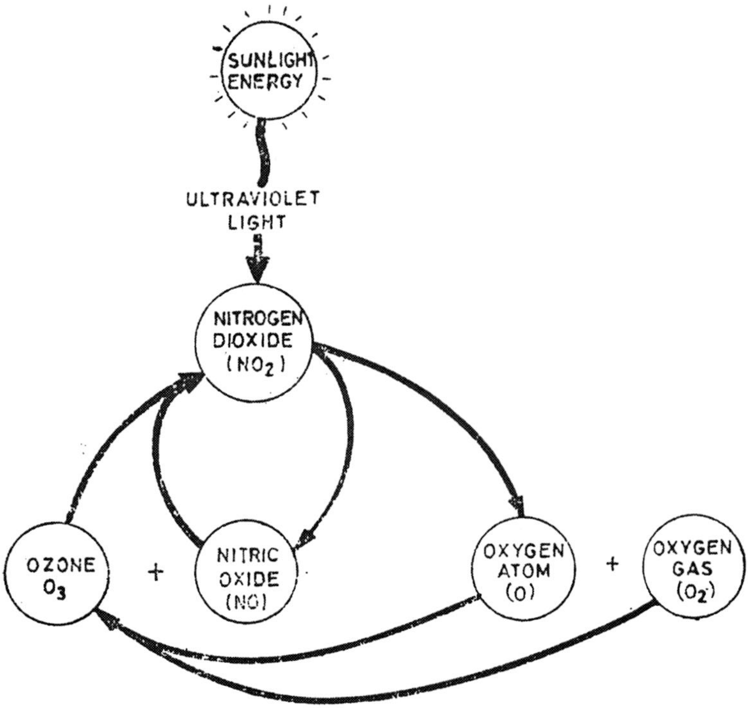

Fig. 6.3. Atmospheric nitrogen dioxide photolytic cycle. The ultraviolet light acts upon nitrogen dioxide, breaking its bond and yielding two substances, nitric oxide (NO) and oxygen atom (O). The oxygen atom then reacts with oxygen (O_2) in the atmosphere, forming ozone (O_3). The ozone reacts with the nitric oxide to reform nitrogen dioxide. Thus the energy of ultraviolet light acts as a "pump" in the rapid destruction and reformation of nitrogen dioxide.

process industries emit nitrogen dioxide into the atmosphere, but in most metropolitan areas the most important source is vehicular exhaust.

Nitrogen dioxide is responsible for much of the ozone in polluted atmospheres due to its decomposition by the action of ultraviolet light:

$$\underset{\text{nitrogen dioxide}}{NO_2} + \underset{\text{ultraviolet light}}{UV} \rightarrow \underset{\text{nitric oxide}}{NO} + \underset{\text{atomic oxygen}}{O}$$

$$\underset{\text{atomic oxygen}}{O} + \underset{\text{molecular oxygen}}{O_2} \rightarrow \underset{\text{ozone}}{O_3}$$

These reactions, in conjunction with others of a complex nature, account for daily cyclical fluctuations in the concentration of ozone in polluted areas of high sunlight intensity. One such area is the Los Angeles basin where the amount of ozone is characteristically low in the morning but rises sharply by afternoon.

Nitrogen dioxide itself is a deadly poison and is one of the most treacherous gases known. Exposure may go unnoticed until it is too late for the victim to recover. Inflammation of the lungs is immediate, but it may not cause enough pain to be noticed. Oedema (accumulation of fluid) may occur several days later, resulting in death. In some cases the effects may be prolonged for weeks or even months. Apollo astronauts Thomas P. Stafford, Donald K. Slayton, and Vance D. Brand narrowly escaped death while making a landing after their historic rendezvous in space with a Soviet spacecraft in July 1975. They were hospitalized after inhaling a poisonous gas, believed to be nitrogen tetroxide, N_2O_4, that leaked into the spacecraft from the fuel system. Much of the nitrogen dioxide in the atmosphere exists in the form of nitrogen tetroxide, with which it is in equilibrium.

$$2NO_2 \rightleftharpoons N_2O_4$$

Even short exposure to nitrogen dioxide is dangerous. Breathing air containing 20 ppm of the gas for a brief time may be fatal. With prolonged exposure, much smaller amounts in the air are dangerous. The MAC (maximum allowable concentration) for occupational exposure has been set at 5 ppm (9 milligrams per cubic metre) for an 8-hour period.

Experiments with rats showed that the acute toxicity of this gas is less than that of ozone, but greater than that of carbon monoxide. However, delayed, rather than immediate, death is the characteristic response of nitrogen dioxide exposure. It is predicted that the nitrogen dioxide concentration will continue to get worse for several years until adequate controls can be established. Nitrogen dioxide remains one of the most potentially dangerous of modern-day pollutants.

Tobacco smoke

Cigarette and cigar smoke contains 330 to 1,500 ppm nitrogen oxides which are removed completely by inhalation, presumably by absorption in the lungs.

Nitrogen dioxide reacts with water to form nitric acid and nitric oxide:

$$\underset{\text{nitrogen dioxide}}{3NO_2} + \underset{\text{water}}{H_2O} \rightarrow \underset{\text{nitric acid}}{2HNO_3} + \underset{\text{nitric oxide}}{NO}$$

Much of the nitric acid in the atmosphere ultimately combines with ammonia to form ammonium nitrate, and probably all of the nitrogen oxides that do not react photochemically become nitrate salts that form aerosols (suspensions of small particles) in the atmosphere in sizes larger than 1 micron.

Hydrocarbons

Hydrocarbons are organic compounds consisting of hydrogen and carbon. They comprise the major portions of coal, petroleum, petroleum derivatives, and similar products. The smallest hydrocarbon is methane, a molecule consisting of 1 carbon atom and 4 hydrogen atoms.

Methane CH_4, is a colourless, odourless, tasteless, nearly nonpoisonous, flammable gas. It is a dangerous asphyxiant because it is heavier than air and displaces oxygen at or near ground level. At high concentrations in the absence of oxygen, it is a narcotic (producing sleep). The gas is widely distributed in nature. Methane is sometimes produced by the decay of organic material in swampy locations, hence the common designation "marsh gas." It is a constituent of natural gas, comprising an average of about 85 per cent of the product. Methane is highly flammable, making it one of the more useful fuels, but it is explosive in concentrations above 5.53 per cent in a mixture with air.

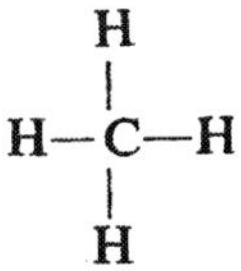

or CH_4

methane

Methane burns with a pale, faintly luminous flame. It can be used as a substitute for petrol in internal combustion engines, and if it were widely adopted, the major exhaust pollutants would be greatly reduced. However, problems of transport, storage, supply, and mileage performance would greatly increase the cost. Known reserves of natural gas are extremely limited and there is increasing industrial demand for methane in the chemical processing industries.

The second smallest hydrocarbon, *ethylene* (C_2H_4), occurs in illuminating gas in concentrations up to 4 per cent. It is also produced naturally by ripening fruit and by certain microorganisms, and in the absence of ventilation it is highly toxic to plants, causing defoliation at concentrations of a few parts per billion. Ethylene has been used as an inhalation anaesthetic but is dangerous because of its high flammability and explosiveness in mixture with air. It is only slightly toxic to higher animals, the lethal concentration for mice being 950,000 ppm.

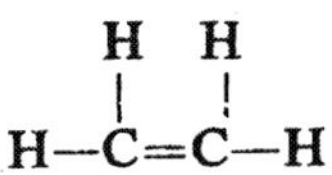

or $CH_2{=}CH_2$

ethylene

There is little promise that pure ethylene, or its relative, *ethane* (CH_3CH_3), can be used extensively for fuel. However, the higher molecular weight members of the family, *propane* and *butane*, with 3 and 4 carbon atoms respectively, are well known as 'LPG', or liquified petroleum. Internal combustion engines operate efficiently with these gases used as fuel, but availability and problems of supply are limitations to their wide-spread adoption.

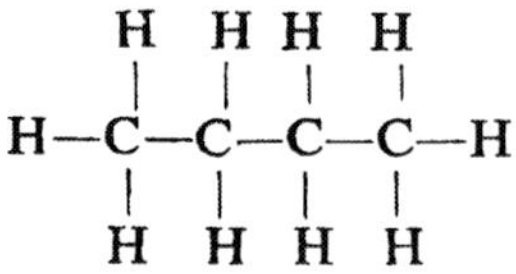

or $CH_3CH_2CH_2CH_3$

normal butane

Petrol is a complex mixture of hydrocarbons containing from 4 to about 12 carbon atoms. The components differ in boiling point from about 38° to 200°C, referred to as the product's boiling range.

The phenomenal success of petrol as fuel for the internal combustion engine is based on the fact that at the very high temperatures that occur in the engine cylinder, oxygen combines violently with the hydrocarbon vapours, causing rapid expansion while converting the reactants to carbon dioxide, carbon monoxide, and water, as well as other end products.

Sources of hydrocarbon pollution

More than half of the atmospheric hydrocarbon pollution comes from automobiles. Much of it consists of unburned petrol in the exhaust from incomplete combustion, although some of the hydrocarbons in automobile exhaust are reaction products that are not found in petrol. The second largest source of hydrocarbons is from industry, accounting for about one-fourth of the hydrocarbon pollution. Evaporation from refineries and storage tanks is an important source. Hydrocarbons derived from petroleum are used extensively in industry as solvents and cleaners. Paint thinner is an example of a mixture of volatile hydrocarbons. The total hydrocabon emission contributes to pollution directly and to smog indirectly by the reactions between hydrocarbons and other constituents in the atmosphere (see Photochemical Smog).

Fuel Additives

Tetraethyl lead

Numerous substances can be added to petrol to inhibit the explosion-type combustion that causes knocking but the metal alkyls (a class of hydrocarbon combined with a metal) are the most effective. *Tetraethyl lead* is the best. But when used alone, tetraethyl lead has a serious disadvantage. A decomposition product of combustion is lead oxide, which is deposited in the combustion chamber and further reduced to metallic lead, which causes pitting. To correct this, petrol manufacturers add *ethylene dibromide* or a mixture of ethylene dibromide and ethylene dichloride, which form lead halides (lead bromide and lead chloride) instead of the oxide. The lead halides are more resistant to reduction than the oxide and are therefore blown out with the exhaust in the form of particulate aerosols.

Lead Pollution of Air

Lead is the most important particulate matter resulting from automotive pollution. There is a direct correlation between petrol consumption and particulate lead in the air. The danger to human health and the environment is discussed elsewhere.

Smog

The word *smog* is supposed to have originated in Great Britain as a short-form of "smoke-fog." The term may have been suggested by a report on smoke-fog deaths by H. A. Des Voeux in 1911. The report dealt with two occasion during the autumn of 1909 in Glasgow, Scotland, when 1,063 deaths were attributed to smoke and fog. In recent years, atmospheric pollution has become increasingly critical. There have been several catastrophic instances of smog resulting in serious injury and multiple deaths.

Killer Smogs

From time to time a combination of high emission of air pollutants and adverse atmospheric conditions combine to cause the formation of "killer smogs". During a disastrous week in the winter of 1930 in the Meuse Valley of Belgium, a large number of people were made sick and many were killed by smog (Firket, 1931). The Meuse Valley is a narrow river valley, 5 kilo metres long with hills on each side. The area is highly industrialized with blast furnaces of steel mills, zinc smelting plants, glass factories, and sulphuric acid plants. On December 1, 1930, and for the rest of the week, a thermal inversion layer confined the increasingly heavy concentration of industrial emissions to the valley. By the third day many people complained of respiratery ailments, and before the atmospheric conditions changed to clear the smog from the valley, 60 people died. There were also fatalities in cattle. It could not be determined with certainty what toxicants caused the difficulty, but probably a combination of several pollutants was responsible. It was estimated that the sulphur dioxide content of the air was 9.6 to 38.4 ppm which if true would be very high indeed (Firket, 1936). If the sulphur dioxide had been completely oxidized to sulphuric acid (although this is improbable), the concentration of sulphuric acid mist theoretically could have reached 14.6 to 58.5 ppm.

The London Smog

The December 5-9, 1952, London fog is the best known air pollution episode in England. It began on Thursday, December 4, as a high-pressure air mass created a temperature inversion over southern England. A white fog formed in the London area. As particulate and sulphur dioxide levels built up, because of extensive use of coal as fuel for space heating and electric production, the fog became a black fog. At the same time the high-pressure area stalled and became stationary. The buildup of pollutants combined with the fog resulted in essentially zero visibility. By Saturday pollutants were sufficiently concentrated to cause deaths, which were estimated to be about 4000. The fog cleared on December 9 (Wise, 1971).

Since 1952, similar conditions have occurred in London in January 1955, December 1956, January 1956, December 1957, January 1959, December 1962 (Table 7.1). However, the smoke levels were markedly lower in 1962 because of the clean up effected by the British Clean Air Act.

Other Air Pollution Episodes

New York

Several episodes of air pollution have occurred in New York which have been responsible for excess deaths, for example, November, 1953 (Greenburg, 1962) and November, 1962 (McCarrol and Bradely, 1966). The 1966 Thanks-giving weekend episode is probably the best known. The maximum 24 hour average of hourly SO_2 values was 0.51 ppm on November 23. On the 24th the value was 0.47 ppm, and on the 25th it was 0.41 ppm. The maximum *hourly* concentration was 1.02 ppm on November 25. There were 168 deaths, mostly older people. The temperature was unusually high during this period. However, temperature effects were not the cause of the deaths.

Bhopal Gas Leak

The world had never heard of Bhopal, a town with a population of just under one million people, located in central India and the capital city of the State of Madhya Pradesh, until the morning of the 3rd of December 1984. From that day onwards for at least a week, Bhopal dominated the front pages of the press world-wide and also received extensive television and radio coverage. Ironically, the Bhopal disaster occurred towards the end of the Golden Jubilee celebrations of Union Carbide India Limited. Today, almost everyone everywhere knows what Bhopal is, where it is, and what happened there.

The Bhopal nightmare (Kharbanda, 1985) happened on the night of December 3, 1984, but it was no nightmare. It was real. Methyl isocyanate (MIC) gas leaked from three storage tanks of the Union Carbide Factory exposing thousands of inhabitants. It was the worst ever industrial disaster. The number of dead (estimates range from 2,000 to over 5,000 but the precise number is not known) exceeds the deaths reported for ALL the industrial accidents in the past forty years. The nightmare was, mercifully, a brief one for the dead. But for the survivors who were exposed to the deadly MIC gas, the nightmare may be lifelong and even beyond because of the serious genetic effects.

It is really ironic to think that the solution was so simple. A balloon to show the wind direction and the simple precaution of breathing through a wet towel could have saved hundreds of lives.

Methyl isocyanate is an intermediate used in the manufacture of carbamate type of pesticides. Carbon monoxide, CO, obtained by partial oxidation of coal is combined with chlorine gas in presence of activated carbon to form phosgene, $COCl_2$. Phosgene is a poisonous gas and was used as a chemical weapon in World War I. Phosgene and methyl amine combine to form methyl isocyanate, the product (MIC) is stored in tanks for further use to produce the carbamate insecticide.

$$\underset{\text{phosgene}}{COCl_2} + \underset{\text{methyl amine}}{CH_3NH_2} \rightarrow \underset{\text{Methyl isocyanate}}{CH_3NCO} + \underset{\text{hydrochloric acid}}{2HCl}$$

Inhalation of MIC causes burning sensation in the eyes, removes oxygen from the lungs and death due to choking. The

Date	*Location*	*Pollutants*	*Effects*
December 1-5, 1930	Meuse Valley, Belgium	Particulates, SO_3 (9-38 ppm)	63 excess deaths, cough, chest pain, eye and nasal irritation—all age groups
December 5-9, 1952	London, UK	Particulates (4000 $\mu g\ m^{-3}$) SO_2 (1.3 ppm)	4000 excess deaths
December 5-10, 1961	London, UK	SO_2 (2.0 ppm)	700 excess deaths
January 29-February 12, 1963	New Yo.k, USA	Particulates, SO_2 (0.5 ppm)	200-400 excess deaths
November 24-30, 1966	New York, USA	Particulates SO_2 (1.02 ppm)	168 excess deaths
December 3rd, 1984	Bhopal, India	Methyl Isocyanate (MIC)	10000 excess deaths, Blindness all age groups

threshold limiting value (TLV) for MIC for man is 0.02 ppm and the LD_{50} (LD = lethal dose) is 2 ppm in air.

The Bhopal catastrophe is certainly not merely a case of the proverbial 'Once-in-a-100 years' event. It was not a freak accident. The Bhopal plant was unprofitable, mainly due to poor capacity utilization. In fact discussions had taken place before December 3rd 1984 on dismantling the plant (Economic Times, 1985). But this never happened.

Another Bhopal—Never Again! is the wording of a resolution passed by the International Organization of Consumers Union (IOCU) at its World Congress in Bangkok just a few days after the Bhopal tragedy. This surely represents the sentiments of all right thinking people everywhere.

Photochemical Smog

The most irritating, and some of the most injurious, components of smog are the products of reactions in the atmosphere between oxygen, ozone, and emission pollutants. The mixture of undesirable products is sometimes called *photochemical smog* because some of the chemical reactions are initiated by the energy from ultraviolet light. The cycle of reactions is not completely understood; much of the work on air pollution is done in *smog chambers*, which may differ in their characteristics from conditions in the atmosphere. However, the important steps in the atmospheric production of photochemical smog have been determined. They consist of a series of reactions in which oxygen, ozone, nitrogen oxides, and hydrocarbons participate to produce compounds that are both irritating and toxic. Some of the principal reactions are as follows:

Step 1

One of the most important steps in the formation of photochemical smog is the action of ultraviolet light on nitrogen dioxide, producing nitric oxide and atomic oxygen:

$$NO_2 \xrightarrow{\text{UV light}} NO + O$$

nitrogen dioxide   nitric oxide   atomic oxygen

Step 2

The atomic oxygen further reacts with a molecule of oxygen with the formation of ozone:

$$O + O_2 \rightarrow O_3$$

atomic oxygen   molecular oxygen   ozone

Some of the ozone is used up in oxidising nitric oxide to nitrogen dioxide (Fig. 6.4).

Step 3

If hydrocarbons are present, they can react with the ozone to form hydrocarbon free radicals which are very reactive because of the unpaired electrons which they contain. These free radicals can react with each other, with atmospheric toxicants and with other components of air to produce photochemical smog:

$$O_3 + RCH_3 \longrightarrow \dot{R}CH_2$$

ozone   hydrocarbon   free radical

$$O + RC\dot{H}_2 \longrightarrow R\dot{C}O$$

atomic oxygen   free radical   acyl radical

The acyl radicals undergo a complex series of reactions, some of which consist of a further reaction with oxygen to form more highly reactive substances called *peroxy radicals*, which in turn react with hydrocarbons to from aldehydes, ketones, and other compounds.

$$R\dot{C}O + O_2 \longrightarrow R\dot{C}OO_2$$

peroxyacyl radical

Step 4

A crucial stage in the photochemical smog complex is the reactionbetween peroxy radicals and nitrogen dioxide to form the highly toxic PANs (peroxyacyl nitrates):

$$\underset{\text{peroxyacyl radical}}{R\dot{C}OO_2} + \underset{\text{nitrogen dioxide}}{NO_2} \rightarrow \underset{\text{PAN (peroxyacyl nitrate)}}{RCOO_2NO_2}$$

The foregoing scheme is a greatly abbreviated and simplified version of the scheme for the formation of the photochemical smog.

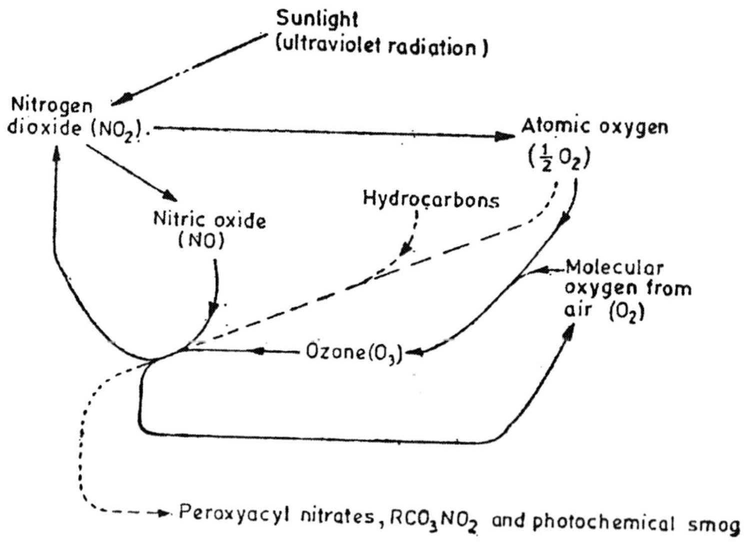

Fig. 6.4. The nitrogen dioxide photolytic cycle, showing the interaction of hydrocarbons to produce photochemical smog.

Hydrocarbons vary widely in their potential to form free radicals. Therefore, the nature of a hydrocarbon mixture must be known before its contribution to photochemical smog can be assessed. In general, unsaturated compounds ($-C=C-$) are the most reactive. Following reactions with ozone, atomic oxygen, molecular oxygen and nitrogen dioxide, they give rise to the group of toxicants known as peroxyacyl nitrates. These toxicants, like ozone, can react further with hydrocarbons to generate free radicals.

PANs

The PANs (peroxyacyl nitrates) are believed to be responsible for most of the eye irritation in photochemical smog, although other compounds are known to contribute. All of the substances responsible for irritation have not been identified. Formaldehyde and acrolein, both highly toxic substances, are known to be present along with various types of peroxides and free radicals.

The first known member of the group is *peroxyacetyl nitrate*, or simply PAN.

$$CH_3-\overset{\overset{\large O}{\|}}{C}-OONO_2$$

It is a potent eye irritant at concentrations of about 1 ppm or less. PANs and other eye irritants are known to persist for more than 24 hours in contrast to ozone, which is believed to have, in photochemical smog, an average half-life of about 1 hour. One distant cousin of the PAN family, PBzN (peroxybenzoyl nitrate), causes eye irritation at a concentration only 1/200 that of formaldehyde. Since it takes approximately twice as much formaldehyde as PAN to produce irritation, the results indicate that PBzN is 100 times more powerful than the familiar PAN.

Oxidant

It is common practice to express the amount of irritating air pollution as parts per million of ***oxidants***. The most potent oxidants in photochemical smog include nitrogen dioxide, PANs (peroxyacyl nitrates), and ozone. These were formerly referred to as a group and called *oxidant*. More recently it is customary to report nitrogen dioxide separately.

Ozone may comprise as much as three-fourths of the total oxidant in irritating smog. Ozone, itself, does not account for the degree of eye irritation experienced. But the concentration of more highly irritating substances is sometimes expressed in terms of the equivalent of ozone in oxidizing power. The first, second, and third alert stages for ozone are set at 0.5, 1.0, and 1.5 ppm, respectively.

Forest Haze

Ozone is a contributor to the hazy atmosphere frequently seen in forested mountain areas. Terpene hydrocarbons (a class of unsaturated organic compounds) are emitted by vegetation. The volatility and concentration of these compounds is such that in calm weather there may be significant concentrations in the atmosphere in and above wooded areas. F.W. Went and R. Rasmussen made a detailed study of the emanations from vegetation. They believe that the terpenes undergo photochemical polymerization to an atmospheric aerosol (finely divided droplets of particulate matter), resulting in the "blue haze" seen in many forested areas. Went estimated that the total release of terpene-type hydrocarbons from vegetation is about 170 million tons per year. Ozone in the atmosphere is probably responsible for most of the terpene reactions that cause the bluish smoke. Thus, a natural haze or "smog" is formed that may be seen especially on warm, calm days in forested or other heavily vegetated areas.

Aeroallergens

Aeroallergens are airborne pollutants, primarily pollen and spores. They usually range from 2 to 50 micrometres diameter. Their major characteristics is that they cause allergic reactions in humans (*e.g.*, allergic rhinitis, often misnamed "hay fever"). To act as an aeroallergen, pollen of a given species must occur in the atmosphere in large quantities close to centers of human population and possess strong antigens that are capable of sensitizing and causing allergy symptoms (Sheldon, Lovell, and Mathews, 1967). Some of the most prevalent pollen aeroallergens come from members of the grass family (*Poaceae*) and from ragweed (*Ambrosia* spp.) and walnut (*Juglans* spp.). Pollen from many other plant species may also act as aeroallergens. Among the fungi, spores of only a few dozen genera have been implicated as aeroallergens. However, these are quite common genera that produce immense quantities of spores, often outnumbering airborne pollen. In the Midwest and northeastern United States, the most important fungal genera appear to include *Cladosporium* and *Alternaria*, spores of *Helminthosporium*, *Aspergillus*, *Penicillium*, *Fusarium*, and *Rhizopus* are of secondary importance (Sheldon, Lovell, and Mathews, 1967).

Other air pollutants, which are more dangerous but do not appear in as great a volume, include asbestos, beryllium, and mercury. These are discussed elsewhere in the book, The US Environmental Protection Agency, EPA, has designated these three substances as hazardous air pollutants.

REFERENCES

Bowonder, B., *et. al.*, 1985. Avoiding Future Bhopals. *Environment*. 27, 6-13.

Carbide Considered Removal before Accident. *Economic Times*, 29th September 1985.

Firket, J., 1931. The Cause of the Symptoms Found in the Meuse Valley during the Fog of December, 1930. *Bill. Roy. Acad. Med.* (Belgium). 11. 683-739.

Firket, J., 1936. Fog Along the Meuse Valley, *Trans. Faraday Soc.*, 32 1192-1197.

Greenburg, L.C., Erhardt, F. Field, J.I. Reed, and N.S. Seriff. 1962. Intermittent Air Pollution Episodes in New York City. *Public Health Repts.*, (U.S.) 78. 1051-1065, 1963.

International Organization of Consumers Unions. 1985. The Lessons of Bhopal—A Community actions resources manual on hazardous technologies. Penang, Malaysia, August 1985, pp. 151.

Kharbanda, O.P. 1985. The Bhopal Nightmare. *Chem. Econ. and Eng. Rev.* 17. 40-41.

Kharbanda, O.P., 1985, Another Bhopal? Never Again. *The Chem. Engr.* (UK) July/August 1985, p. 40.

McCarrol, J. and W. Bradley. 1966. Excess Mortality as an Indicator of Health Effects of Air Pollution. *Am. J. Public Health.* 56, pp. 1933-1942.

McGraw-Hill Encyclopedia of Environmental Sciences, 1974, McGraw-Hill Book Company, New York.

Sheldon, J.M., R.C. Lovell, and K.P. Mathews. 1967. *A Manual of Clinical Allergy*. 2nd ed. W.B. Saunders Company, Philadelphia.

The Chemical Industry After Bhopal. An International Symposium held in London on 7/8th Nov. 1985, IBC Technical Services Ltd., London.

Wagner, R.H., 1978. *Environment and Man.* 3rd, ed., W.W. Norton & Company, Inc., New York.

Wise, W., 1971. *Killer Smog.* Audubon-Ballantine Books, New York.

7

Ozone Depletion

A proportion of the Sun's energy is emitted as ultraviolet (UV) radiation or light. Ultraviolet light is divided into three types according to wavelength :

UVA 320–400 nm

UVB 290—320 nm

UVC $<$ 290 nm

DNA, the genetic code present in all living cells, is damaged by ultraviolet radiation. UVC is the most damaging radiation followed by UVB, and UVA. Damage to DNA can either kill the cell or cause it to mutate and become cancerous.

In the stratosphere there is a layer of ozone, O_3, which plays a critical role in protecting living organisms from the harmful effects of ultraviolet radiation. This layer is concentrated at 20 to 25 kilometres from the surface of the earth. Ozone is broken down by sunlight, and this process absorbs shorter wavelength ultraviolet light (ultraviolet radiation below 315 nm). Stratospheric ozone therefore acts as a shield around the Earth, which virtually eliminates UVC and greatly reduces the amount of UVB reaching the surface of the planet. Without the protective effect of ozone, terrestrial ecosystems could not have evolved in the manner that they have.

Pollutants Destroying the Ozone Layer

Ozone is simultaneously formed and destroyed in the stratosphere by naturally occurring chemical reactions. When oxygen molecules absorb sunlight (light energy), they dissociate into free oxygen atoms which then combine with O_2 to form ozone :

$$O_2 + h\nu \rightarrow 2O$$

$$O + O_2 \rightarrow O_3$$

Sunlight is also responsible for the destruction of ozone. It dissociates the ozone molecule into oxygen atoms and molecules. The free oxygen combines with ozone to form more oxygen molecules.

$$O_3 + h\nu \rightarrow O + O_2$$

$$O + O_3 \rightarrow 2O_2$$

SSTs and Nuclear War

In the early 1970s, during the investigations into possible environmental consequences of a fleet of supersonic transports (SSTs), it was discovered that nitrogen oxides (NO_x) emitted in the exhaust fumes of the aircraft could accelerate the ozone layer's breakdown:

$$NO + O_3 \rightarrow NO_2 + O_2$$

H. Johnston at the University of California, Berkeley in 1971 estimated that a fleet of SSTs would emit enough NO_x into the stratosphere to pose a threat to the ozone layer.

Further research and exhaustive reviews by panels of experts under the auspices of the National Academy of Sciences and the United State Department of Transportation completely substantiated Johnston's work.

Nuclear bombs offer an additional avenue by which significant amounts of nitrogen oxides can be added to the stratosphere. The high temperature in nuclear fireballs produces nitrogen oxides in abundance from atmospheric nitrogen and oxygen. Much of that material is deposited in the stratosphere. Although there is some controversy about the issue, several investigators believe that ozone depletion by atmospheric bomb tests conducted in the early 1960s was great enough (around 5 per cent) to show up in the relatively crude historical data that are available for stratospheric ozone concentrations.

Chlorofluorocarbons, CFCs

F.S. Rowland and M.J. Molino (1974) at the University of California, Irvine, reported that common aerosols were a far more dangerous threat to the ozone layer than the nitrogen oxides. These chlorofluorocarbon compounds commonly known as Freons, break down in the stratosphere and produce chlorine atoms, which are six times more effective than the nitrogen oxides in destroying the ozone layer.

Chlorofluorocarbons (CFCs) are man-made gases which are relatively inexpensive to produce, non-flammable, non-toxic, non-reactive and very stable. Since their invention in the early 1930s they have been developed for use as propellants in aerosol cans, as blowing agents to aid cell production in plastic foams (particularly polyurethane and polystyrene), as refrigerants in refrigerating and air conditioning equipment and as solvent cleaners for use particularly in the micro-electronics industry.

CFCs are simple compounds produced from hydrocarbons by replacing one or more of the hydrogen atoms by both chlorine and fluorine atoms. When all the hydrogen atoms have been replaced, the CFC is said to be fully halogenated; partially halogenated compounds (HCFCs) still contain one or more hydrogen atoms. Six grades are produced in significant quantities : the fully halogenated CFCs 11, 12, 113, 114 and 115, and the partially halogenated HCFC 22. Another group of gases, known as Halons, are derived from CFCs by the substitution of one or more bromine atoms into the basic CFC molecules. Halons, referred to by the numbers 1211, 1301 and 2402, are used mostly in fire extinguishers. At present, worldwide annual demand for all types of CFCs exceeds 1 million metric tonnes.

The majority of CFC production is located in the European countries, USA and Japan although in recent years manufacturing plants have been built in India, South Africa, Australia and several countries in Central and South America, Eastern Europe and the Far East.

CFCs are emitted into the atmosphere either directly from aerosols; through leakage from refrigeration units or air conditioning systems; by escape from factories during the foam-blowing process; and at the end of the waste stream, for example when refrigerators are dumped. Concern first developed in the early 1970s when it was realised that CFCs emitted into the atmosphere would not break down rapidly in the way that most chemicals do, but could persist for several decades.

The CFCs are popular propellants because they are easily liquefied at relatively mild pressures, and they vapourize readily at room temperatures. It is their stability that makes them a threat to the ozone layer. Once they are released, they gradually ascend unchanged through the troposphere, perhaps over a period of years, until they reach the stratosphere. There, the strong ultraviolet light from the sun causes them to dissociate

$$CF_2Cl_2 + h\nu \rightarrow CF_2Cl + Cl$$
$$CFCl_3 + h\nu \rightarrow CFCl_2 + Cl$$

The Cl atom then causes the following chain of reactions, which leads to the breakdown of the ozone layer (Bartell and Ritz, 1974) :

$$Cl + O_3 \rightarrow ClO + O_2$$
$$ClO + O \rightarrow Cl + O_2$$
$$\overline{O_3 + O \rightarrow 2O_2 \text{ (net)}}$$

Studies by R.J. Cicerone *et. al.*, in 1975 at the University of Michigan, revealed that the CFCs can destroy the ozone layer to the extent of 5 per cent by the end of the century if we do not increase our use of CFCs. This could reach 40 per cent if we increase our use of CFCs at the rate we have been increasing them in the past few years.

Measurements made from 1971 to 1974 show that there has been a definite increase in CFCs concentration in the troposphere (Wilkniss et al., 1975).

Further studies were made by several scientific groups. The results of their investigations indicated that the CFCs deplete the ozone layer.

Nitrogen Fertilizer

The man-made threats to ozone discussed so far—SSTs, nuclear bombs, and aerosol propellants and refrigerants—share the characteristic that human well-being does not depend in any fundamental sense on the use of the offending products.[34] SSTs are not economically necessary, harmless substitutes can be found for most applications of chlorofluoromethanes and it should be plain that nuclear bombs are a product the world can do without. Thus, the future magnitude of those threats could be limited without serious social cost merely by not producing or using them any longer. An intrinsically more difficult problem is posed by the possibility that nitrogen fertilizer is the source of yet another threat to ozone.

The danger is that nitrous oxide (N_2O) produced from nitrates and nitrites in soil or water by denitrifying bacteria is oxidized in the upper atmosphere to ozone-destroying nitric oxide. Applying more nitrogen fertilizer may add mere nitrous oxide to the atmosphere, which means more nitric oxide in the stratosphere, which means less ozone. This threat to the ozone is less certain than the others discussed here, because of uncertainties in the atmospheric chemistry of N_2O and in the magnitude of flows in the natural nitrogen cycle. Some models of the processes at work suggest that ozone depletion from this cause is likely to be small and gradual.

If the worst proves to be true, however, forecasted increases in the use of nitrogen fertilizer could produce a *15-per cent reduction* in stratospheric ozone by the year 2025.

Other Threats to Ozone

There may be other compounds as dangerous to the ozone layer as those already mentioned, Carbon tetrachloride (CCl_4), for one, is added to the lower atmosphere by processes both natural and human. There is evidence that human activity is the principal source, and that a significant quantity of the carbon tetrachloride reaches the stratosphere. It is conceivable that chlorination of drinking water and of sewage are important sources that have some effect on ozone. Hammond (1975) reported that bromine also acts to break down the ozone layer in a manner similar to chlorine. Methyl bromide (CH_3Br), widely used to fumigate agricultural land, is thought to be the source of stratospheric bromine.

The Antarctic Ozone "Hole"

According to the Environment Committee Report (London, 1988) a rapid, and increasingly severe, depletion of the ozone layer over Antarctica has occurred between late August and mid-October each year since the late 1970s. This depletion occurs at the polar vortex, the unique meteorological phenomenon of an isolated air mass with very low temperatures surrounded by the Roaring Forties and described as "like a chemical containment vessel". The British Antarctic Survey observed in September and October 1987 the most severe depletion yet seen there.

In two months in 1987, 50 per cent of the total column of ozone was lost; and in a 5 km region of the atmosphere, from 15 km to 20 km, 97 per cent of the ozone was lost.

The Natural Environment Research Council (NERC) of England reported that the ozone "hole" has deepened by 1987 to an extent which would be considered catastrophic if it occurred over temperate or tropical regions, because the increase in ultraviolet radiation would be damaging to all life forms.

Important evidence began to emerge in the USA from the Ozone Trends Panel which was set up in October 1986 by the National Aeronautics and Space Administration (NASA). The Panel undertook a critical re-analysis and interpretation of all available measurements of total column ozone, and of vertical profiles of ozone. In the case of the ozone 'hole' in the second half of August and September 1987 two aircraft were flown repeatedly into the Antarctic vortex equipped with instruments to measure the chemical composition and meteorological state of the atmosphere. Preliminary results from the Panel confirm the extent of ozone depletion in the polar vortex.

In addition, they report that :

(i) the total columns of ozone in spring time, 1987, at all latitudes south of 60 degrees south, were the lowest since measurements began 30 years ago;

(ii) in 1987, the region of low column ozone persisted until late November/early December, which is the longest since the region of low ozone was first detected;

(iii) while the column ozone depletion is largest in the Antarctic spring, ozone appears to have decreased since 1979 by 5 per cent or more at all latitudes south of 60 degrees south throughout the year.

Global Depletion of Ozone

The US Environmental Protection Agency's (EPA) unpublished data from equipment aboard the Nimbus-7 satellite show that the decrease in total column ozone was about 4 per cent between 1978 and 1985. The most up-to-date information is now emerging from the Ozone Trends Panel in the USA. The Panel examined and re-examined every available measurement of column and vertical ozone and reported that there has been a 3 per cent decrease in annual average ozone since 1969. The study also shows that the January mean column ozone between latitudes 53 degrees north and 64 degrees north (which in Britain includes everywhere north of Nottingham) has decreased by 8 per cent since 1969. The fears of the NASA scientists are that a 'hole' similar to that which appears in the Antarctic could also occur in the Arctic.

Are CFCs Responsible for the Depletion of the Ozone Layer?

In August 1987 the UK Stratospheric Ozone Review Group, in a report prepared at the request of the DOE and the Meteorological Office, referring to the Antarctic ozone "hole" stated that the cause is not yet known. It may be a natural change in atmospheric circulation; it may be chemical, due to increasing levels of chlorine and/or bromine from man's activities or it may involve both of these.

The Environment Committee is of the opinion that man-made chlorine compounds are primarily responsible for the observed rapid decrease of ozone during September within the polar vortex.

The committee in its report (1988) recommended that the Department of Environment (DOE) should take a strong lead in co-ordinating European research into ozone layer depletion and monitoring and that appropriate Government funds be made available for this purpose.

The Committee further recommended that the DOE increase its support to the CFC manufacturers' research programme into the environmental effects of ozone depletion, as well as academic and other independent research.

The Montreal Protocol

The discovery of the ozone hole over Antarctica has prompted political action. The Vienna Convention for the protection of the ozone layer was established in 1985 under the auspices of the United Nations Environment Programme (UNEP). In September 1987, the world's prime consumers and manufacturers of CFCs and halons reached an agreement in Montreal to reduce emissions into the atmosphere. The Montreal Protocol covers the main halogenated CFCs and halons of commercial significance. These are listed in Table 7.1 along with their chemical names, their ozone depletion potential, and their 1985 global production rates.

The Montreal Protocol signatory nations agreed to a freeze on the consumption and production of group 2 substances at their 1986 level. For group 1 substances the Protocol requires a freeze on consumption at 1986 levels by 1990, a cut to 80% by 1994, and a further cut to 50% by 1999. Production, on the other hand, is allowed to rise by 10% by 1990 before reducing to 90% by 1994 and 65% by 1999. The reason for the

TABLE 7.1. **Chemicals Included in the Montreal Protocol**

Substance		*Ozone depletion potential**	*1986 world production (million kg/year)*
Group 1			
CFC 11	Trichlorofluoromethane	1.0	370
CFC 12	Dichlorodifluoromethane	1.0	470
CFC 113	Trichlorotrifluoromethane	0.8	160
CFC 114	Dichlorotetrafluoroethane	1·0	Very low
CFC 115	Chloropentafluoroethane	0.6	Very low
Group 2			
Halon 1211	Bromochlorodifluoromethane	2.7	10
Halon 1301	Kromotrifluoromethane	11.4	10
Halon 2492	Dibromoterafluoroethane	5.6	Very low

* Data from "*Ozone Depletion*"—*Health and Environmental Consequences*. (1989) Ed. Robin Russel Jones and Tom Wigley. John Wiley and Sons. England.

difference between consumption and production is to make allowance for developing and low-consumer countries who might need to import CFCs.

Many developing countries including India did not sign the Protocol. India releases just 6000 tonnes of CFC's a year which is equivalent to one and a half days release of World's Total.

Consequences of Ozone Depletion

The existing ozone layer screens out more than 99 per cent of incoming solar energy in the wavelength region between 230 and 320 nm (nanometre); radiation of those wavelengths is harmful to most forms of terrestrial life, and the small fraction that gets through is known to cause sunburn, skin cancer, and eye damage in human beings, as well as various kinds of damage to other animals and plants. Most scientists agree that terrestrial life did not evolve until the ozone layer was established. The special effectiveness of ultraviolet radiation in damaging living systems is due to the sensitivity of the genetic material DNA and of proteins to those wavelengths. DNA exhibits a resonance—the capacity to absorb radiant energy very efficiently—at wavelengths around 265 nm. Proteins typically have resonances in the range of 275 to 285 nm. The energy absorbed when radiation of those wavelengths strikes the DNA and protein molecules damages them, causing the many disruptive effects now known, and probably others as well.

Analytic and computer models of varying degree of sophistication have been used to estimate what increase in the ultraviolet radiation reaching the surface would be associated with different degree of depletion of the stratospheric ozone. Although the relation between ozone depletion and increased ultraviolet radiation at the surface varies with latitude and other factors, a rough average is that a 5-per cent decrease in ozone produces a 10-per cent increase in radiation. The best-studied effect of such an increase is an increase in the incidence of human skin cancers, the magnitude of which is estimated to be about equal to the percentage change in radiation. The effect of ultraviolet radiation on the human skin is a cumulative process; that is, malignancies may be induced with repeated exposure to ultraviolet radiation. Studies made with mice indicate that ultraviolet wavelengths in the range 280 to 320 nm are most effective in tumour development. Since loss of stratospheric ozone would lead to increased intensities in precisely this range, it is clear that increased occurrence of human skin cancer could result from destruction of the ozone layer.

It has been estimated that a 5-per cent decrease in ozone and a 10-per cent increase in ultraviolet radiation would produce 20,000 to 60,000 additional skin cancer per year in the United States alone, and perhaps 5 times as many worldwide. Incidence of skin cancer is higher among light-skinned people than among dark-skinned. About 5 per cent of the skin cancers would be fatal. About 30-per cent reduction in ozone would be expected to produce on the order of 1 million additional skin cancers per year worldwide, of which perhaps 50,000 would be fatal.

Although an increase in skin cancer is the best understood of the effects of ozone depletion, it is by no means the only and not necessarily even the most serious effect. Laboratory tests have indicated that increased ultraviolet radiation impairs the growth of certain crop plants and adversely affects a wide variety of organisms from bacteria to vertebrates. Phytoplankton, the primary producers at the base of oceanic food webs may prove especially vulnerable if the light-receptors that trigger the protective dive response to excess radiation are not sufficiently sensitive to ultraviolet light. Furthermore, many insects "see" ultraviolet light and rely on it for navigation; an increase in ultraviolet light may interfere with their pollination function.

Acceleration of the mutation rate of microorganisms because of DNA's sensitivity to ultraviolet light could have adveres effects on human beings through disease, causing more frequent development of strains of microorganisms against which people would have no resistance and of strains resistant to antibiotics. The consequences for ecosystems of differential susceptibility among different organisms to increased ultraviolet radiation cannot be predicted with any accuracy, but such differential susceptibility is known to exist and the possibility of its resulting in serious instability cannot be excluded.

Nonbiological damage from increased ultraviolet radiation is likely to include faster deterioration of paints, fabrics, plastics, and other materials. The magnitude and economic cost of such effects is uncertain—and dependent on the amount of increased radiation—but it would surely be significant.

The impact of ozone depletion on the climate is probably the least certain. The situation is extremely complicated: ozone absorbs and reradiates radiant energy in the infrared as well as the ultraviolet part of the spectrum of wavelengths; so do products of some of the reactions that consume ozone, and so do most of the catalysts for the ozone-destroying reactions and their precursors (chlorine is such a catalyst, the chlorofluorocarbons are its precursors). One recent study based on are latively simple model indicates that continued growth of chlorofluorocarbon production could produce a significant global warming by the year 2000, independent of the effect on ozone, merely because of a greenhouse effect in which those compounds would block outgoing infrared radiation. Changes in the radiation absorbing and emitting properties of the atmosphere not only change the direct radiation flows at Earth's surface, however, they also change the vertical temperature distribution in the troposphere and the stratosphere, which in turn changes the rate of vertical circulation. All of this could influence climate.

REFERENCES

Bartell, L.S., and C.L. Ritz. 1974. Stratospheric ozone destruction by man-made chlorofluoromethanes. *Science* 185, 1163-1164.

Cicerone, R. J., S. Walters, and R. S. Stolarski. 1975. Chlorine compounds and stratospheric ozone. *Science* **188**, 378-379.

Hammond, A. L. 1975. Ozone destruction : Problem's scope grows, its urgency recedes. *Science* 187 (4182), 1181—1183).

Jone, Robin Russel and Tom M. L. Wigly (Eds.). 1989. *Ozone Depletion: Health and Environment Consequences.* John Wiley & Sons. England.

The Environment Committee Report on Air Pollution. 1988. Vol I. London. Her Majesty's Stationary Office.

Wilkniss, P. E., J. W. Swinnerton, R. A. Lamontagne, and D. J. Bresson. 1975. 'Trichlorofluoromethane in the troposphere—Distribution and increase', 1971—1974. *Science* **187**, 832-833.

8

Atmospheric Ecology and Air Pollution

The Weather and Our Atmosphere

Most life on earth is dependent on green plants, which in turn depend on the weather for their growth. The food we eat, the type and location of the house we build, the fuel we burn, the clothes we wear, as well as our health, recreation, and transportation are all influenced by weather and climate. Although the terms "weather" and "climate" are often used interchangeably, there is a significant difference in their meaning. *Weather* is defined as the state of the atmosphere with respect to such things as temperature, humidity, wind, cloudiness, and precipitation. *Climate* is defined as the characteristic weather conditions of a given place averaged over a long period of time.

An atmosphere is necessary as a medium for weather. The atmosphere is an envelope of air surrounding the earth. Besides supplying air, it protects living things from the harmful radiation of the sun and outer space. The atmosphere is composed of a mixture of gases whose percentage remains relatively constant. Its principal components are nitrogen, 78.1 per cent, oxygen 20.9 per cent and argon, 0.9 per cent. It also contains traces of neon, helium, krypton, sulphur dioxide, nitrogen oxides, and varying amounts of water vapour, carbon dioxide, and pollutants.

Fig. 8.1 shows the average temperature and heights of the lower layers of the atmosphere. Although the atmosphere extends upward hundreds of kilometres, until it meets the rarified interplanetary medium of the solar system, 99 per cent of its mass is found within 30 kilometres of the earth's surface. Most of our weather occurs in the troposphere, which extends an average of 12 kilometres above the earth's surface and contains 80 per cent of the atmospheric total mass. The outer edge of the troposphere is called the *tropopause*. Its height

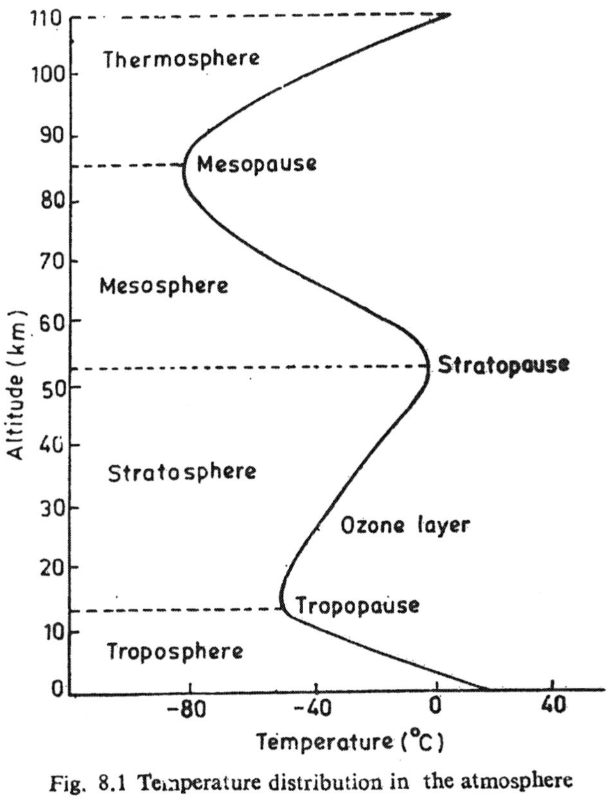

Fig. 8.1 Temperature distribution in the atmosphere

varies from an average of 16 kilometres at the equator to 8 kilometres at the poles. The next layer, the stratosphere contains the ozone layer whose destruction has become a topic of great concern.

In order to have weather, there must also be heat which is provided by the sun. Some of the sun's heat bounces back from the earth's atmosphere, some is reradiated by the earth, some warms the earth and its atmosphere, and some provides motive force for the global weather and energy to change water into water vapour. Heat provides the convective force that lifts masses of air vertically. Since the sun shines more directly on the equator than on the poles, it heats the earth's surface in the equatorial regions more than in the polar areas. Solar heating and radiational cooling set up temperature differentials between the equator and poles. Temperature differences occur between oceans and land masses, because bodies of water gain and lose heat more slowly than land. Temperature differentials lead to pressure differentials, since cold air is heavier than warm air. Pressure differences generate air movement or winds. Air always moves from high to low pressure, causing both the soft, gentle sea breeze and the violent winds of a tornado. With reference to a global scale, heat increases the buoyancy of the air near the equator. Since there is a horizontal temperature and pressure gradient due to the unequal heating between the equator and the poles, the rising air tends to move towards the poles and the cold polar air moves toward the equator to fill the vacated space. These wind movements are not directly north and south, because they are greatly affected by friction and the earth's rotation. This deflection of the winds to the right in the northern hemisphere and to the left in the southern hemisphere is known as the *Coriolis force*. The general global wind and pressure patterns thus formed are shown in Fig. 2.4 (see Chapter 2). This global wind pattern exists mostly in theory. Many forces are superimposed on it, to change the wind's directions, especially in the temperate zones. These forces include land and sea breezes, day and night unevenness of heating, ocean currents, the uneven surface of the land, regional wind systems such as the monsoons, and the huge migrating air masses with their leading edges known as *fronts*. If a cold mass of air advances toward warm air, it will push itself under the warm air, causing the latter to rise abruptly, forming a line of cumulus-type clouds with frequent showers and thundershowers. This is known as a *cold front*. Squall lines and tornadoes are frequently associated with the cold front when the temperature contrast between the two air massess is extreme and the warm air contains much moisture. When a warm mass of air advances, it moves slowly over the cold air and forms a wide band of stratus-type clouds. This is a *warm front*. It is accompanied by a steady type of precipitation and, frequently, fog. When a front shows little movement, it is called a *stationary front* and usually causes cloudiness and precipitation, often of a several days' duration.

Another essential ingredient of weather is moistu.e. The weather we notice as we look out of the window—cloudes, fog, rain, snow, or sleet—is some form of water. Moisture exists in the air as invisible water vapour until it is cooled below its saturation point and condenses out on tiny particles to form water droplets or ice crystals. Thus clouds are formed. If a cloud rests on the ground, it is known as *fog*. The water droplets in a cloud may combine to form larger drops, which finally fall as *rain*. If the rain drops fall through a layer of freezing air, they reach the ground as *sleet*. Ice crystals falling from the clouds may reach the earth as *snow* if they pass through cold air. If they pass through warm air, they may reach the ground as rain. The vertical currents in a thunderstorm cause ice particles being formed in the cloud to rise and fall many times alternately through freezing levels and warm moist ones until masses form of sufficient size to fall out of the cloud as *hailstones*.

In recent years, there has been an increasing interest in what effect human activity might have on the changes in world climate. Twenty thousand years ago, ice covered the places where many of the world's largest cities now stand. Some of that ice still remains in the polar regions and in the Antarctic. It is estimated that if the Greenland ice cap alone were to melt at a speed comparable with that reached following the last ice age, the sea level would be raised about 7.6 metres (25 feet), coastal regions would be inundated, and much of our present land would be covered with water.

In the past, climate had been generally considered to be relatively static, except on the geological time scale. Recently, it was noted that a definite warming trend existed from the mid-nineteenth century until 1940, when quite an abrupt cooling trend started. A closer study revealed alternate cooling and warming trends since the time of recorded history. A brief outline of the warmer and cooler periods is given below:

8000-7000 B.C.	The last major ice sheet disappeared from Scandinavia
5000-3000 B.C.	Postglacial climatic optimum (the world temperature was 2 to 3°C higher than now)
900-450 B.C.	Decline from the climatic optimum
1000-1200 A.D.	Secondary optimum of climate
1430-1850	Little Ice Age
1850-1940	Warming trend
1940-	Cooling trend

The reasons for these worldwide temperature fluctuations remains a matter of conjecture. Some think the cooling trends are due to less radiation from the sun, caused possibly by sun spots. Other theories suggest that an increase of volcanic dust caused more reflection of the sun's radiation back to space, thus causing the earth's cooling. Air pollution has been blamed for recent climate change. The warming trend since 1880 parallels the great increase in the use of fossil fuels, which was accompanied by an increase in CO_2. This suggests that the warmer global temperatures were caused by the *greenhouse effect* of CO_2.

The Greenhouse Effect

Much of the sun's energy is emitted as heat rays, consisting of radiation in the infrared portion of the electromagnetic spectrum. The temperature at the surface of the earth is determined by the energy balance between the sun's rays that strike the planet and the heat that is radiated back into space. The near-infrared rays of sunlight penetrate the earth's atmosphere relatively unimpeded, and some of the heat is absorbed and retained by the earth or objects on the surface. The heated earth then reradiates this absorbed energy as radiations of longer wavelength, mainly in the middle-infrared portion of the spectrum. Much of this does not pass through the air envelope to outer space but is absorbed by the carbon dioxide and water vapour in the atmosphere and adds to the heat that is already present. Though carbon dioxide is almost completely transparent to visible light, and partially to near-infrared, it strongly absorbs and reradiates the rays of longer wavelength, especially heat rays having wavelengths of 12×10^{-6} to 18×10^{-6} m. Thus carbon dioxide acts like the glass of a greenhouse, and on a global scale, tends to warm the air in the lower levels of the atmosphere. This is called the *greenhouse effect* (Fig. 8.2).

Water vapour and ozone are also absorbers of infrared radiation and help to keep the earth warm. Water vapour is an efficient absorber of infrared of around 15×10^{-6} m and of another band of wavelengths around 6.3×10^{-6} m. Ozone, an absorber of infrared of around 9.6×10^{-6} m, is relatively minor in its effect.

Weather and Air Pollution

Even though it is not certain that pollution affects global climate, it does affect local climate. A city tends to be warmer, cloudier, rainier, and has lower visibility than the surrounding

(a) (b)

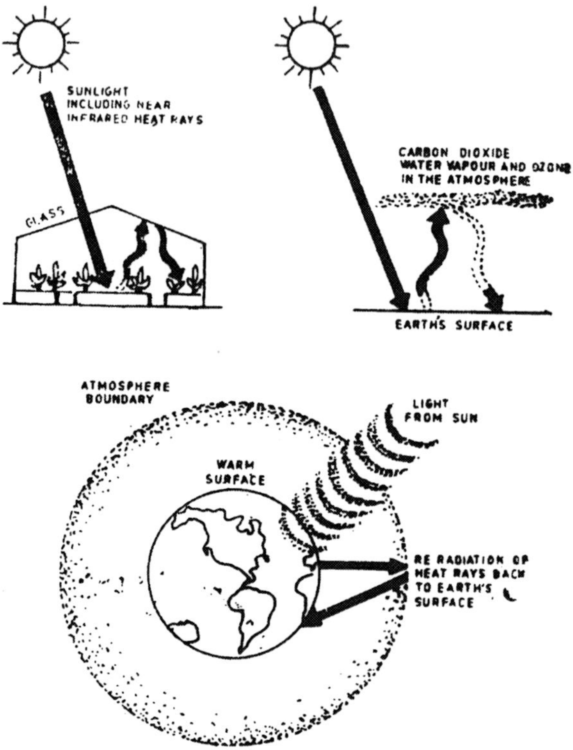

(c)

Fig. 8.2 Greenhouse effect. (a) Heat from the sun readily penetrates the glass roof of a greenhouse in the form of near-infrared rays. These are absorbed by soil, plants, and other objects and reradiated as longer wavelength heat rays. However, the glass is not as transparent to these radiations of longer wavelength; thus much of the heat is held inside the greenhouse. (b) In a similar way, the atmospheric carbon dioxide, water, and ozone are transparent to the near-infrared rays of the sun but absorb the reflected longer wavelength heat rays, thus (c) holding the heat within the boundaries of the earth's atmosphere and having a warming effect.

countryside. The first day frost comes later in the city. Concrete walls and streets absorb heat during the day and then radiate it at night. Industries also help to increase a city's temperature.

Although heat produced by combustion does have a local effect, especially in the heat islands forming over cities, it appears to have no global significance. Fears have been expressed about the depletion of oxygen because its consumption in the combustion of fuels is thought to be greater than its supply from photosynthesis. Recent studies have found no evidence that the concentration of O_2 is changing. However, if the phytoplankton that supplies most of our O_2 is destroyed by either water pollution or an increase in ultraviolet radiation, we may have real reason for concern.

Weather determines what happens to pollutants after they reach the atmosphere. Surface winds help scatter pollutants; the greater the speed, the better the dispersal. Convective currents also help. During daylight hours, the earth's surface is warmed by the sun, causing currents of warm air to rise upward, carrying pollutants with them. As the warm air rises, it cools, because the pressure on it is decreased, allowing it to expand and cool. It continues upward until it becomes the same temperature as the surrounding air. The distance between the earth and this altitude is known as the *mixing depth*. It determines the depth to which pollutants will be mixed. Its height is continually changing, and it is higher in the summer than in the winter.

Temperature Inversion

In the troposphere, air normally decreases in temperature with an increase in altitude. However, during a temperature inversion, a layer of warm air lies over the cooler air and acts as a lid, keeping the smoke and other pollutants down near the earth's surface. Temperature inversions occur under several different meteorological conditions. When a warm front moves over static cooler air, a temperature inversion is formed. On clear cool nights, the ground radiates its heat skyward and cools the air at low levels while warmer air prevails aloft. The disastrous smog episode in London in December 1952 was due to a third condition, which is caused by subsidence. This inversion results when upper layers of air descend or subside during a developing anticyclone or high-pressure area. The air warms as it descends, and the warming is greater at higher levels than near the ground. This results in a temperature inversion. If the anticyclone remains stationary, the inversion may persist and the pollutants accumulate until severe conditions develop. Perhaps the most widely recognized temperature inversions are those that occur over Los Angeles. They usually occur on at least 300 days during a year. These inversions are caused by local meteorological conditions augmented by the mountainous topography of the area. The cold Humboldt ocean current flows along the California coastline west of Los Angeles. As the prevailing westerlies with their warm moist air from the South Pacific pass over this current, the lower air layers are cooled, often forming fog, whereas the upper layers remain warm. Thus, when the air moves eastward over Los Angeles, which is surrounded by mountains, the cooler lower air layers with fog and the city's air pollutants become trapped not only by the lid of warm air above it but also by the surrounding mountains.

The Climate Debate

In this final decade of the 20th century the greenhouse effect and the prospects of global warming is the subject of scientific and political controversy and a cause for widespread concern. A quantitative assessment of global warming trends can be made on the basis of historical temperature records. Unfortunately, such records are difficult to come by. Today global temperature measurements are compiled through the World Weather Watch System, a global cooperative network of national meteorological services. Observations in previous centuries were largely compiled by individual observers working without coordination. About 10 years ago, in the face of growing global climate concerns, Phillip D. Jones and Tom M.L. Wigley and coworkers at the Climatic Research Unit of the University of East Anglia, Norwich, England, initiated a project to collect and analyze, once and for all, every available historical temperature record. This was sponsored by the

U. S. Department of Energy. Raymond S. Bradley of the University of Massachusetts at Amherst and Henry F. Diaz of the Environmental Resources Laboratory of the National Oceanic and Atmospheric Administration (NOAA); USA were their collaborators. Their assessement is that the earth has experienced an overall warming trend of half a degree celcius since the late 19th century.

The relevance to climate of the condition of Earth's surface is obvious. The most important features are: (1) presence or absence of water, (2) reflectivity, or albedo, (3) ability to transfer water to the atmosphere, (4) capacity to store heat, and (5) topography and texture. These interrelated features determine how much of the incident radiant energy is captured, how it is distributed between surface and atmosphere, how the surface interacts with the winds, and how the shoreline interacts with ocean currents. Civilization can influence the surface conditions by changing or removing vegetation, by damming and diverting rivers to form reservoirs and irrigation projects, by "reclaiming" land from shallow seas as bays, by covering land with pavement and buildings, by spilling oil on water, snow, and ice, and by releasing previously stored energy as heat.

The earliest significant human influence on climate was undoubtedly that of vegetation changes brought about by human use of fire, deforestation, cultivation, and grazing. Subsequently, changes brought about by irrigation projects joined this category of climatic impacts associated with the provision of food and fiber.

As far as is known, the actual changes in climate associated with such surface alterations have been mainly regional, rather than global. One reason for this may be that some human activities, such as irrigation of arid lands, actually decrease albedo (reflectivity) which compensates to some degree for man-made increases in the albedo elsewhere.

Regional changes can themselves be devastating. There has been particular interest in possible climatological explanations of the prolonged drought of the late 1960s and early 1970s in sub-Saharan Africa.

The activities of civilization certainly have modified Earth's surface significantly, and they appear to be modifying the stratosphere. What of the troposphere, where most of the mass of the atmosphere resides and most of the motion that "produces" climate takes place ? Apart from the influences already mentioned, civilization's direct impact on the troposphere is mainly of three kinds: an increase in the concentration of carbon dioxide, which is worldwide and uniformly distributed; an increase in the concentration of particles in the atmosphere over large regions (but little if any increase in other regions); and the production of high-altitude clouds initiated as jet aircraft contrails at the tropopause. The principal impact of all these phenomena occurs through the interaction of the contaminants with incoming solar radiation and outgoing terrestrial radiation.

Carbon Dioxide

Beginning in 1957, the Mauna Loa Climate Observatory at Hawaii, USA collected data which revealed a systematic increase in atmospheric carbon dioxide. To date, the change from 290 parts per million (ppm) in 1880 to 352 parts per million in 1989 represents more than a 20 per cent increase over the course of the past century. Efforts to unravel the climatic consequences of increasing carbon dioxide emissions have been going on since long. The world famous mathematician John Von Neumann at the Institute for Advanced study in Princeton, N.J. (USA) made the first attempt to represent the atmosphere mathematically on digital computers in 1950. In 1963, an unusual laboratory of the National Oceanic and Atmospheric Administration (NOAA) was established at Princeton University under the leadership of Joseph Smagorinsky. The laboratory was devoted to the mathematical modeling of the atmosphere using the largest and the fastest digital computers available. Called the Geophysical Fluid Dynamics Laboratory, the center kept researchers from many nations interested in this new approach to the study of the atmosphere. Among them was a young Japanese scientist, Syukuro Manabe who developed the first climate model in collaboration with his colleague Richard T. Wetherald in the 1960's. In 1975 they calculated that a doubling of the carbon dioxide content of the atmosphere would produce a global climate warming of about 3°C averaged over the surface of the earth. This calculation has been verified in many different laboratories and has not changed substantially. Studies have since been undertaken in many parts of the world, including Europe and the Soviet Union. In the U.S. the National Research Council conducted studies in 1966, 1977, 1979, 1983 and 1987.

Although the mathematical models of all the groups yielded similar results, the details of the geographic distribution of climate changes differed from one model to the other. All projected that an increase in carbon dioxide would bring about a gradual warming, but the timing of this warming would depend on the rate of global energy use. They all agreed that if reasonable assumptions were made about future global energy consumption, it would be around the middle of the next century that the carbon dioxide content of the atmosphere might double.

Just how much this doubling of carbon dioxide would increase temperatures, however, varied greatly from model to model. Some predicted as little as 1°C increase, whereas others predicted increases of as much as 5°C.

These projected temperature changes may appear innocuous because variations of this magnitude are experienced in the normal course of daily and seasonal weather. Their full implications can be appreciated by noting that it took only 1°C average decrease in temperatures in Europe to cause the run of several frigid centuries (from the 1400's to the 1800's).

known as the Little Ice Age. Five degrees C is believed to be the difference in temperature that separates the end of the last great ice age 12,000 years ago from the present. Further, the projections (Fig. 8.3, 8.4 and 8.5) indicate that the Northern Hemisphere would experience in just a half century an unprecedented temperature change, 10 to 50 times faster than the change since the last ice age.

Since the carbon dioxide content of the atmosphere has increased by more than 20 per cent over the past century, we ought to be able to detect the climate warming in the global temperature record during the same period. Researchers have sought to do this, but it is a much more difficult task than it first appears. The problem is that climate is always in a state of natural fluctuation. Separating out the changes that are caused by increasing carbon dioxide from the natural changes is difficult. Moreover, the climatic temperature record is based on scattered and irregular observations not taken specifically for the purposes of determining climatic conditions.

Even so, careful analysis of these temperature records by scientists in the U.S. and in the U.K. sought to detect whether a climate warming has occurred and whether such warming is consistent with the prediction of the models. The prevailing view is that the climatic record over the past century for the entire globe reveals a net increase in temperature ranging from 0.3 to 0.8°C. But set against this conclusion is the disturbing result that similar increases in temperature cannot be detected over the past century in the U.S., where observations are numerous and accurate.

Even if the temperature rise is real, a puzzle remains that workers have not been able to solve: Is the rise in global temperatures a natural fluctuation or a result of the increase in greenhouse gases ? All that can be said is that the observed increase is consistent with the lower end of the temperature increases predicted by the computer models. Consequently, the temperature records, as well as the predictions of mathematical

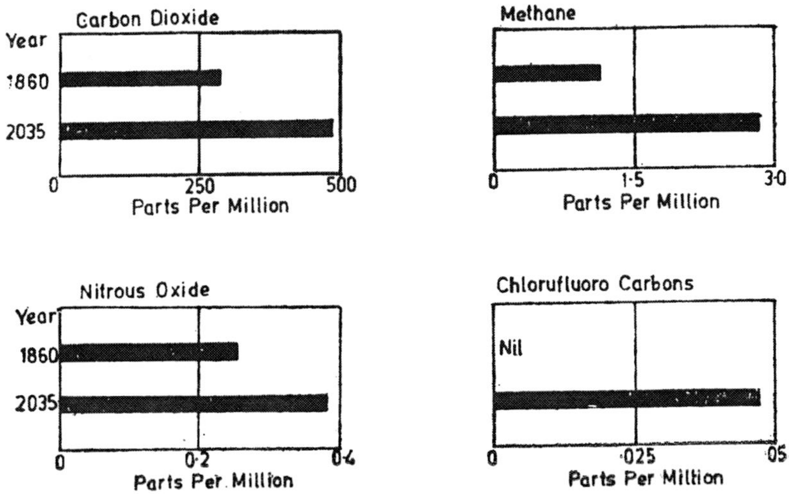

Fig 8.3 The projected growth of greenhouse gases in the atmosphere.

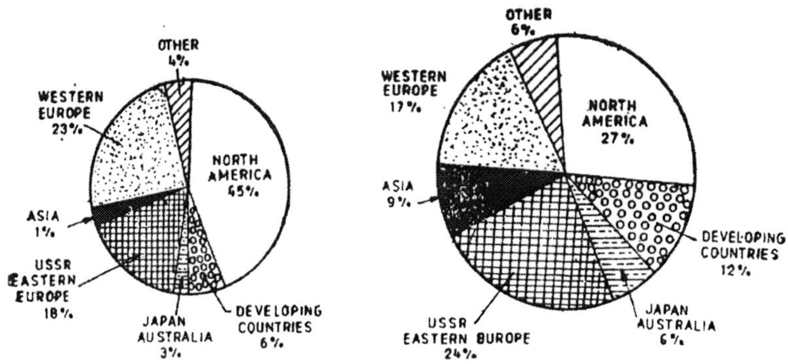

1950
1618 million tons

1980
5170 million tons

Fig. 8.4 The changing pattern of global carbon dioxide emissions

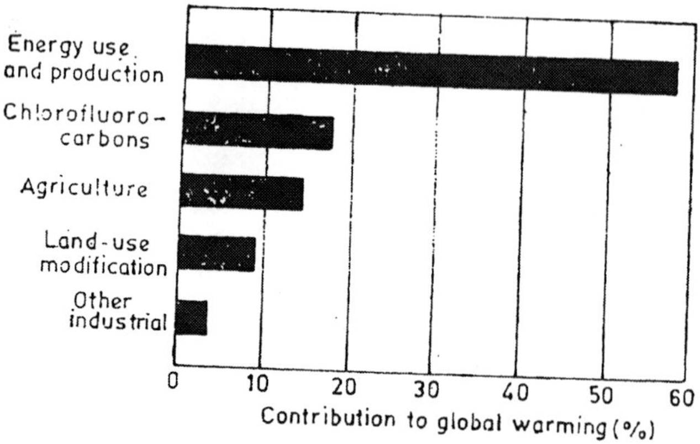

Fig. 8.5 Human activities that may cause global warming

models, provides support to both viewpoints: those who believe the evidence warrants action now and to those who believe the evidence is still too weak.

Particles

Particles in the atmosphere are of many sizes and compositions, depending on their origins: smoke from forest fires, agricultural burning, and the combustion of fossil fuels; solid and liquid particles that began as gaseous emissions from combustion and natural processes; sea salt; dust from volcanic explosions and from agricultural activity.

Unlike carbon dioxide, particles generally do not become uniformly mixed in the atmosphere; their life expectancy before falling out is too short. Therefore, the concentration of particles in the atmosphere tends to be high for some distance downwind of the major sources and low elsewhere. The smallest particles remain in the atmosphere longest and therefore have the biggest effects; this means that sources of small particles, whose contribution to emissions may not be very large by weight, make a disproportionate contribution in terms of climatic impact.

There is no doubt that major volcanic eruptions, when they occur, are the single greatest source of atmospheric particles; such eruptions put a considerable quantity of debris into the stratosphere, where it remains long enough to spread over an entire hemisphere. Climatic records indicate that uncommonly cold years have followed some of the major volcanic eruptions of the past 200 years, presumably because the particles have screened out some sunlight.

Particles play an important role in climatic processes by serving as condensation centres for the formation of water droplets or ice crystals when the relative humidity exceeds 100 per cent. In general, there is virtually always a surplus of atmospheric particles that could serve as condensation nuclei if the relative humidity were very high. However, the actual dynamics of cloud formation is governed by that class of particles whose properties permit condensation when the relative humidity first barely exceeds 100 per cent. That class of particles, called *cloud nuclei*, is defined as particles that are active condensation centres below 101 per cent relative humidity. Particles that have the property of initiating freezing in a droplet that contains them are called *ice nuclei*. Under many circumstances, an increase in the atmospheric concentration of cloud nuclei can increase cloud cover but decrease rainfall, because more but smaller droplets are formed, and small droplets do not fall as readily as large ones. Man-made particles can also have the effects of increasing rainfall and decreasing cloud cover—this is what is done intentionally in cloudseeding programmes to stimulate rainfall—but those phenomena seem to be rarer than those with the opposite effect (decreasing rainfall, increasing cloud cover).

The term *aerosol* is often used interchangealy with *particles* in the literature of climate, although technically *aerosol* refers to particles of less than 1 micron in diameter together with the gas in which they are suspended.

The Consequences

If climate change is a historical fact and a future inevitability, why worry ? The argument is often put forward that people have survived climate change in the past and will continue to do so. Moreover, if civilization adds its own influences to natural climatic trends, how do we know the result won't be an improvement ?

The response to this has several parts. First, to talk merely of survival as a species is to evade the issue of the degree of social disruption and human suffering that have accompanied regional climatic change in recent history and, presumably, at times in the more distant past.

Today, moreover, the absolute magnitude of the human misery that could be caused by climate change is much larger than it ever was because the human populations in virtually all regions of the world are much larger than they were before. Far more people would be affected by any given regional climatic change, and there is not likely to be any room in adjacent regions into which they could flee.

The part of the ecosphere most sensitive to climatic change —that is, the element most easily disrupted to the point of directly influencing human well-being—is agriculture. Farmers know very well that a year's crop can be ruined by rain that comes too early or too late, or in too great quantity or too little, by too much hot weather or too much cold, by early frosts, and by other vagaries of weather.

The consequences of changed climatic conditions can be seen in historical records. One of the recent climatic events was the disastrous drought of the summer of 1988 in USA. During this drought, one of the worst on record, the water in the Mississippi River fell so low that navigation was impossible over long stretches, urban water supplies were threatened and crops throughout the grain belt were devastated. Both officials and the public wondered whether this was the greenhouse effect manifest. Records show that in the U.S. five of the years of the 1980's were among the hottest on record, and the average temperature for the decade as a whole was the warmest since instrumental records have been kept. The temperature went up to 32°C. The worst affected was the corn belt. Leaves of corn plants curled up to conserve water. Pollens missed the flowers and consequently no corn was produced.

In fact we do not know enough to predict the severity of the consequences, because the warming would not be uniform over the surface of the earth. Some parts of the earth would become warmer, some wetter and some drier. On the basis of the evidence at hand climateologists are not able to predict who would benefit and who would lose in such a global redistribution of so-called climatic resources.

Some aspects of global climate warming would be greatly beneficial according to agricultural researchers. Increased carbon dioxide will enhance more active photosynthesis and improve crop growth. About the lowered plant requirements for water in a CO_2-enhanced atmosphere nothing can be said with certainity. Models do, however, agree that the polar regions of the world would undergo greater increases in temperature than would the tropics. Some of the projections of temperature increases in polar areas are startling in their magnitude, predicting as much as 10°C (by 2050) on the average in the Northern Hemisphere and only slight increases in tropical regions. What are the general consequences of such a change in the temperature difference between equatorial and polar regions? If arctic regions were to undergo significantly greater warming than equatorial regions and if precipitation belts were to move farther north, countries in the north temperate and polar zones would probably benefit greatly. Their growing season would lengthen, and their precipitation would increase. With suitable soils, agriculture might thrive. The polar ice caps would melt resulting in rise in sea level by 2 to 3 ft; this would mean loss of coastal land. Low-lying island nations such as the Maldives could be completely wiped out creating the world's first "Greenhouse Refugees". Fires, floods, and prolonged droughts may be more common. These are however, speculations.

Another, somewhat more speculative respect in which climate change could lead to great increases in human misery is by altering the abundance and the geographical distribution of various disease-producing organisms. As is the case with crops, the degree to which such organisms and the other organisms that transport them (vectors) thrive is governed by such environmental conditions as temperature and moisture, in terms of both averages and extremes. Changes in climatic patterns therefore might give certain of those organisms access to human populations that have no prior evolutionary experience with them and hence little or no resistance to them. Alternatively, such changes might remove checks on the abundance of organisms pre-existing in area, to the extent that a previously minor hazard becomes a plague. This is true of pathogens that attack crops and trees, as well as those that attack people.

The Alternatives

If the climate changes, the expectation is that it will do so gradually. We should be able to see the initial evidence of coastal inundation in an increasing frequency of high tides and in the undercutting of seacoasts. Climate warming itself should be evident in a rising frequency of heat waves or in other weather anomalies. The effects of a global climate warming are likely to take 30 to 50 years to become serious, and that is a long enough span in which actions to adapt to these changes should be possible.

Where might we start? According to experts energy conservation and efficiency should be the first priority. Achieving greater energy efficiency justifies itself in economic terms, increased energy efficiency would also reduce urban air pollution and acid precipitation. Shifts in the fossil-fuel mix from coal and oil to natural gas could significantly reduce carbon dioxide emissions per thermal unit. Technology is also available for more efficient power generation and for increased gasoline mileage.

Major investments in non fossil-energy sources are desirable. The circumstances favour significant new investments in possibely safe, publicly acceptable nuclear power. Further development of forms of solar energy—photovoltaics or biomass, for example—makes good sense. Reforestation and forest preservation yields many ecological and climatic benefits. Research aimed at producing stress and disease-resistant crops would also be recommended.

It is likely that humanity will have to adapt to some climate changes. Modes of adaptation by society have not been well studied. Individuals, corporations and communities can adapt to climatic changes in numerous ways. Farmers can change crops, water use can be regulated and management practices can be altered.

Other modes of adaptation would be needed if climate changes were severe. Sea-level rise, which is one of the predicted consequence of a climate warming, might inundate low-lying coastal areas and cause salt water to intrude into freshwater bodies. If this were to occur, society would have to decide whether to invest in protective structures along coasts or adapt by changing land-use patterns. The North Sea dikes in the Netherlands are an outstanding example of adaptation to relative rise in sea level.

The developing nations are not severely affected so far. They can escape the severity of such climatic changes by not adapting the same technology as the Western countries. The choice, may be harder to make such as finding alternatives to freons which might be very expensive. However, a simple check could be to control the population growth.

Our global environment is under attack on many fronts. Climate warming is but one, perhaps the most complex, of these issues. If the changes occuring in our atmosphere are likely to cause consequences, we must understand the problems and promote sensible policies to remedy them.

REFERENCES

'Analysis of Temperature Rords.' Wigley, T.M.L., James K. Angell and P.D. Jones. *in Detecting the Climatic Effects of Increasing Carbon Dioxide*. Ed. M.C. Mac Cracken and F.M. Luther. 1985. U.S. Department of Energy, Carbon dioxide Research Division, DOE/ER-0235.

Bolin, Bert and B.R. Doos, 1986. *The Greenhouse Effect, Climate Change, and Ecosystems*. John Wiley & Sons. Inc, New York.

Jones, P.D., T.M.L. Wigley and P.B. Wright, 'Global Temperature Variations between 1961 to 1984'. *Nature* 322, 430-434.

Jones, P.D. and T.M.L. Wigley, 1990. 'Global Warming Trends.' *Scientific American* 263, 84-91.

Lamb, H.H. 1966. *The Changing Climate*. Methuen and Company Ltd., London.

Lamb, H.H. 1972. *Climate: Present Past, and Future*: Vol. 1, *Fundamentals and Climate Now*. Methuen and Company Ltd., London.

Likens, G.E., and F.H. Bormann. 1974. Acid rain: A serious regional environmental problem. *Science* 184 (4142), 1176-1179.

Mansfield, T.A., ed. 1976. *Effects of Air Pollutants on Plants*. Cambridge University Press, New York.

Schneidar, Stephen H. 1989. The Greenhouse Effect: Science and Policy. *Science* 243, 771-781.

9

Air Pollution and Plants

Almost all air pollutants causing plant injury are gases, but some particulate matter or dusts may also affect vegetation. Some gas contaminants such as ethylene, ammonia, chlorine, and sometimes mercury vapours, exert their injurious effects over limited areas. Most frequently they affect plants or plant products stored in poorly ventilated warehouses in which the pollutants are produced by the plants themselves (ethylene), or from leaks in the cooling system (ammonia).

More serious and widespread damage is caused to plants in the field by chemicals such as ozone, sulphur dioxide, hydrogen fluoride, nitrogen dioxide, peroxyacetyl nitrate (PAN), and particulates. In many localities the air pollutants spread into the area surrounding the source(s) of pollution, become trapped and cause serious plant damage. More frequently, most air pollutants are transported downwind from the urban or industrial centers in which they are produced and may be carried by the wind to areas that are several miles, often hundreds of miles and sometimes thousand of miles, from the source.

How Air Pollutants Affect Plants

For the past 20 years, scientists have been growing plants under controlled conditions and exposing them to various concentrations of common air pollutants. Characteristic injuries to certain plants can be interpreted to show not only the presence, but also the relative concentration, of the air pollutant. Tobacco is an effective ozone indicator. Pinto bean plants are used to detect peroxyacetylnitrate (PAN) and gladiolus to determine fluoride accumulations. Excellent detectors of SO_2 are dahlias, petunias, alfalfa, and cotton. High concentration of or long exposure to these chemicals cause visible and sometimes characteristic symptoms on the affected plants. The leaf is the primary indicator of injurious effects. More important economically, however, is the fact that even when plants are exposed to dosages less than those that cause acute damage, they can still suppress growth, cause dwarfing, early maturation and a decreased crop yield. Moreover, prolonged exposure to air pollutants seems to weaken plants and to predispose them to attack by insects and by some pathogens.

Physiology of the Leaf

Figure 9.1 shows a leaf and its internal tissue. The leaf veins function in the same manner as the blood vessels do in animals, acting as the transport system for water, minerals, and food. The leaf tissue is in layers with a skin or epidermis layer on top and bottom and the photosynthetic cells in between. The stomata are openings in the leaf bottom (and in some leaves in the top) through which CO_2 enters to take part in the photosynthetic process. These openings are protected by two guard cells which open and close to allow gases to enter or leave the leaf. Such gases can, of course, also include pollutants.

Damage to Leaves

Damage to leaves takes several forms. *Necrosis* is a killing of tissue and can result from any of several air pollutants. *Chlorosis* is the loss or reduction of the green plant pigment, chlorophyll, with a resulting yellowing of the leaves. Leaf *abscission* is a dropping of leaves and leaf *epinasty* is a down-

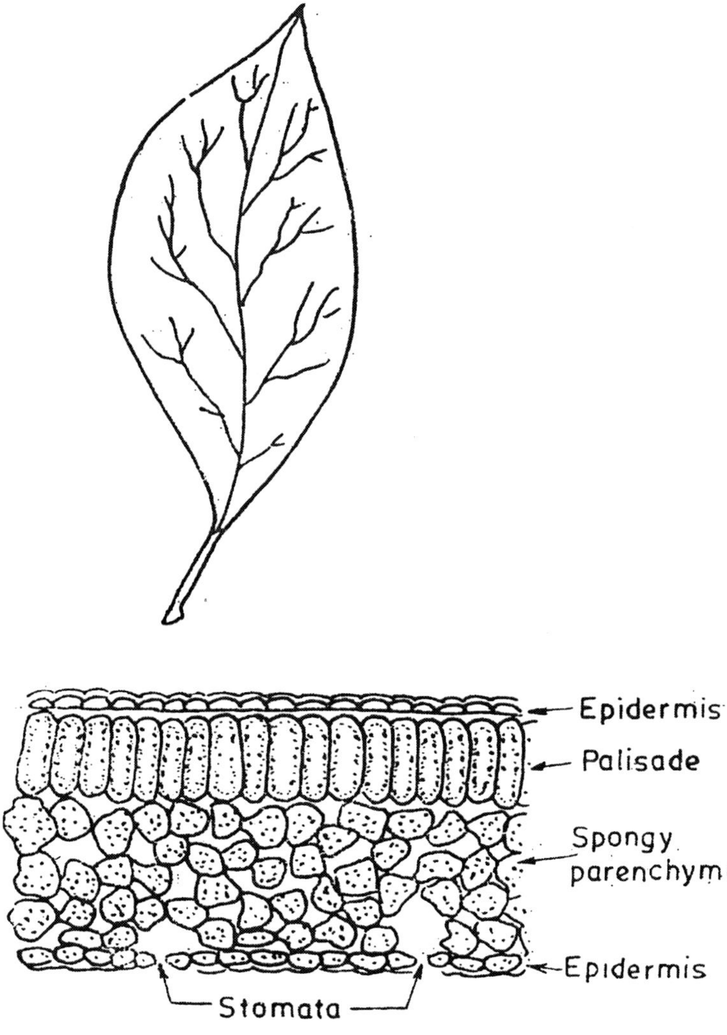

Fig. 9.1 A leaf and its internal tissue.

ward curvature of the leaf due to a higher rate of growth on the upper surface. Plant damaging substances are termed phytotoxicants.

Three kinds of injury are commonly considered important.

1. Acute injury results from short-time exposure to relatively high concentrations, such as might occur under fumigation conditions. The effects are noted within a few hours to a few days and could, for example, be visible markings on the leaves due to a collapse and death of cells. This leads to necrotic patterns, that is, areas of dead tissue.

2. Chronic injury results from long-term low level exposure and usually causes chlorosis or leaf abscission.

3. The third form of injury is an effect on growth without visible markings. Usually a suppression of growth or yield occurs. This is more difficult to document since careful experiments must be done with and without the pollutants under carefully controlled conditions, such as in a greenhouse.

The full effects of chronic injury or growth suppression have been appreciated only recently as it is difficult to separate out the adverse effects of air pollution from other phenomena. Plant diseases may show symptoms which are very similar to those caused by air pollution. Poor plant care or high temperatures may also cause an appearance which is similar to that of a plant damaged by air pollution. Thus in diagnosing the effects of air pollution one must consider such variables as relevant plant disease, nutrition history, insect damage, weather damage, and the type of pollutants in the area.

The effect of concentration versus time is also an important variable in assessing damage. A high dose for a short time may cause an acute injury whereas the same total dose over a longer time may cause no visible effects at all. The concentration-time "law" may be expressed as

$$t(C-C_o)=K$$

where t = time in hours to produce a certain effect on a certain species

C = concentration of a specific gas in parts per million

C_o = the threshold concentration of the gas, in parts per million, to cause injury

K = an experimentally determined constant.

If we rewrite this equation as

$$C = \frac{K}{t} + C_o$$

then plotting C versus $1/t$ gives a straight line whose intercept C_o is the threshold for injury. Specific pollutants and their effects are listed in Table 9.1.

The concentration at which each pollutant causes injury to a plant varies with the plant and even with the age of the plant or the plant part. As the duration that the plant is exposed to the pollutant is increased, damage can be caused by increasingly smaller concentrations of the pollutant until a minimum dose—injury threshold is reached. Plant injury by air pollutants generally increases with increased light intensity, increased soil moisture and air relative humidity, increased temperature, and with the presence of other air pollutants.

Carbon Monoxide

Carbon monoxide has the effect of inhibiting nitrogen fixation. At concentrations of 0.01 to 1 per cent (100 to 10,000 ppm), visible effects are abscission (leaf drop), premature aging, and the initiation of roots on stems. These effects resemble those of ethylene, a gas produced naturally by ripening fruit and some pathogenic fungi. Carbon monoxide is believed to inhibit cellular respiration in plants by reacting with the enzyme system cytochrome-oxidase which is vital to the utilization of sugars as metabolic fuel. In view of this, there has also been interest in the possible effects of carbon monoxide on the cytochrome-oxidase system of animals. Cytochrome oxidase is the principal oxygen-reducing enzyme of the body. It is a heme protein that will combine with carbon monoxide reducing the enzyme activity. However, it appears unlikely that the carbon monoxide concentration becomes high enough to inhibit cytochrome activity except under extreme conditions.

TABLE 9.1. **Air Pollution Injury to Plants**

Pollutant	*Source*	*Susceptible Plants*	*Symptoms*	*Remarks*
1	2	3	4	5
Ozone (O_3)	Automobile exhausts. Other internal combustion engines. (Released NO_2 combines with O_2 in sunlight→O_3) From stratosphere. From lightning, from forests.	Expanding leaves of all plants, especially tobacco, bean, cereals, alfalfa, petunia, pine, citrus, corn.	Stipping, mottling, and chlorosis of leaves, primarily on upper leaf surface. Spots are small to large blenched white to tan brown, or black. Premature defoliation and stunting occurs in plants such as citrus, grapes, and vines.	Enters through stomata. It is the most destructive air pollutant to plants. A major component of smog.
Peroxyacyl nitrates (PAN)	Automobile exhausts or other internal combustion engines. (Gasoline vapours and incompletely burned gasoline+O_3 or NO_2→PAN).	Many kinds of plants, including spinach, petunia, tomato, lettuce, dahlia.	Causes "silver leaf" on plants, *i.e.*, bleached white to bronze spots on lower surface of leaves that may later spread throughout leaf thickness and resemble ozone injury.	Particularly severe near metropolitan areas with smog and inversion layers.
Sulphur dioxide (SO_2)	Stacks of factories. Automobile exhausts, and other internal combustion engines.	Many kinds of plants, including alfalfa, violet, conifers, pea, cotton, bean. Toxic at 0.3 to 0.5 ppm	Low concentration cause general chlorosis. Higher concentrations cause bleaching of interveinal tissues of leaves.	Also combines with moisture and forms toxic acid droplets (acid rain).

Nitrogen dioxide (NO_2)	From oxygen and nitrogen in the air by hot combustion sources, *e.g.*, furnaces, internal combustion engines.	Many kinds of plants including beans, tomatoes. Toxic at 2 to 3 ppm.	Causes bleaching and bronzing of plants similar to that caused by SO_2. At low concentration it also suppresses growth of plants.	
Hydrogen fluoride (HF)	Stacks of factories processing ore or oil.	Many kinds of plants, including corn, peach, tulip. Actively growing, especially wet leaves, are most sensitive. Toxic at 0.1 to 0.2 ppb.	Leaf margins of dicots and leaf tips of monocots turn tan to dark brown, die and may fall from the leaf. Some plants tolerate HF up to 200 ppm.	HF may evaporate or be washed out of plant and plant recovers slowly.
Chlorine (Cl_2) and Hydrogen chloride (HCl)	Refineries, glass factories, incineration of plastics. Near swimming pools and sewage disposal plants.	Many kinds of plants, usually near the source. Toxic at 0.1 ppm.	Leaves show bleached necrotic areas between veins. Leaf margins often appear scorched. Leaves may drop prematurely. Damage resembles that caused by SO_2,	
Ethylene (CH_2CH_2)	Automobile exhausts. Burning of gas, fuel oil, and coal.	Many kinds of plants. Toxic at 0.05 ppm.	Plants remain stunted, their leaves develop abnormally and	Ethylene is a plant hormone with numerous functions.

(Contd.)

TAELE 9.. (*Contd.*)

1	*2*	*3*	*4*	*5*
	From ripening fruits in storage.		senesce prematurely. Plants produce fewer blossoms and fruit. Fruit, *e.g.* apples, develop depressed, necrotic, dark areas (scald).	
Particulate matter (dusts)	Dust from roads, cement factories. Burning of coal, etc.	All plants.	Forms dust or crusty layers on plant surfaces. Plants become chlorotic, grow poorly, and may die. Some dusts are toxic and burn leaf tissues directly or after dissolving in dew or rainwater.	

Nitrogen Dioxide

Nitrogen dioxide is also highly injurious to plants. It suppresses growth when plants are exposed to 0.3 to 0.5 ppm for 10 to 22 days. A concentration of 1 ppm for 8 hours produces significant growth reduction with no visible damage. Exposure to 4 to 8 ppm for 1 to 4 hours causes visible leaf injury to sensitive plants. Low-light intensity increases the plant's sensitivity.

Sulphur Dioxide

Sulphur dioxide has been known for many years to be a major cause of injury to higher plants. Some species are so sensitive that its effects are sometimes the first indication that the air is polluted. The cause of the symptoms, however, is not always easy to identify due to the fact that the injury often resembles other afflictions such as sunscald, frost, aging, or other disorders having obscure environmental or cultural causes.

Plants with high physiological activity are among the most sensitive species. Alfalfa, grains, squash, cotton, grapes, apple, and white pine are susceptible. Species with thick, waxy leaves, such as citrus are more resistant. Trees with needles, such as most pines, are generally resistant except for those with newly formed needles. Conditions that are ordinarily beneficial to plant growth, such as high morning light intensity, high humidity, good soil moisture, and moderate temperature are apt to enhance injury.

Prolonged exposure of plants to sublethal levels of sulphur dioxide causes them to develop a chronic type of injury which first appears between the veins, leaving the vein areas green. Older leaves gradually become yellow, then white, and eventually die. However, in leaves that are only partially injured, the unaffected parts regain their function upon removal from exposure, and new foliage grows normally. The symptoms of such chronic exposure are sometimes described as "early aging" and are not always easy to identify. Concentrations that cause chronic injury are usually less than 0.4 ppm.

Since sulphur dioxide is absorbed through the leaf stomata, conditions that favour or inhibit the opening of stomata similarly affect the amount of sulphur dioxide absorbed. After absorption by the leaf, sulphur dioxide reacts with water and forms phytotoxic sulphite ions. The latter, however, are slowly oxidized in the cell to produce harmless sulphate ions. Thus, if the rate of sulphur dioxide absorption is slow enough, the plant may be able to protect itself from the buildup of phytotoxic sulphites.

The most serious cases of plant damage have been from high concentrations in the vicinity of smelters. A pollution episode occurred at Anaconda, Montana, during 1910 to 1911. Large amounts of sulphur oxides were emitted, causing nearly all the important species of trees within a radius of 8 to 13 kilometres to die. During another incident in 1929 in the area of a smelter at Trail, British Columbia, damage to plants occurred as far as 85 kilometres from the smelter. Douglas fir, ponderosa pine, and other forest vegetation were 60 to 100 per cent damaged for a distance of 50 kilometres. The emission rate from the smelter was as high as 20,000 tons and averaged 18,000 tons of sulphur dioxide per month. Inhibition of cone production by pine trees was still evident two years after installation of a control system. Acute injury to broad-leaved plants causes the leaves, or areas of the leaves, to dry out and become bleached. Injured grasses show a similar pattern except that the tips are killed and the injury generally has a more streaked appearance. The response in plants is clearly the result of the oxidizing or reducing properties of sulphur oxide and not due to the action of acid. Sulphuric acid aerosols are not generally toxic to plants, although emissions of large droplets from factories have been known to damage vegetation.

Sulphur dioxide has a powerful toxic action on micro-organisms. It is used as a fungicidal fumigant, as a preservative for food and wine, and as a disinfectant in breweries. Its toxic action is believed to depend on the formation of sulphurous acid or sulphite ions in the tissues of the organism.

Ozone

Ozone has long been known, to be toxic to vegetation, but injury to crops from atmospheric pollution has only recently been recognized. In 1958 grape leaves in California vineyards were described as having the upper surfaces stippled from ozone toxicity. Shortly afterward similar injury to tobacco leaves, called "weather fleck," was identified as ozone injury. Since then, ozone toxicity has appeared as stipple, flecks, spots, streaks, tipburn, or premature yellowing on at least twenty crop plants. Ozone affects the mature leaves first and spots are usually visible only on the upper surface. White pine needles are also affected with tipburn caused by ozone.

Plants are extremely sensitive to ozone. Concentrations as low as 0.08 ppm for periods of exposure as short as 1 hour cause injury to sensitive plant tissues. The sensitivity of plants to ozone is dependent upon the species, variety, and environmental conditions both before and during exposure. Among the edible plants that are susceptible to ozone, some varieties of tomatoes and radishes are injured by 0.15 to 0.25 ppm when exposed for 1 to 2 hours.

Ozone is taken into leaves through stomata and injures primarily palisade but also other cells by disrupting the cell membrane. Affected cells near stomata collapse and die and appears as white (bleached) necrotic flecking or stippling on broad leaves, as streaks on the leaves of cereal crops, and as browning on the tips of pine needles.

Many crop plants, such as alfalfa, bean, citrus, grape, potato, soybean, tobacco, and wheat, and many ornamentals and trees, such as ash, lilac, several pines, and poplar, are quite sensitive to ozone, while some other crops, such as cabbage, peas, peanuts, and pepper are of intermediate sensitivity, and some, such as beets, cotton, lettuce, strawberry, and apricot, are tolerant. Flecking of tobacco leaves reduces their value, particularly as cigar wrappers.

Certain tobacco varieties are the most sensitive organisms known for indicating the presence of ozone and will show symptoms when the atmospheric concentration is above 0.05 ppm. The presence of sulphur dioxide enhances the effect of ozone. For example, symptoms appeared after 2 hours of exposure of 0.037 ppm ozone plus 0.24 ppm sulphur dioxide, but when the plants were exposed to the same concentrations of the gases separately, they did not develop visible signs of injury. This effect is described technically as reducing the threshold concentration by synergistic action.

Smog

Before the biological importance of air pollution was fully realized, it was noted that plants grown in bad smog areas suffered injury from what often appeared to be obscure causes. It is now apparent that many plants are more sensitive than humans to the acute effect of smog and, in fact, can be used as indicators of harmful levels of air contaminants.

Two ingredients of photochemical smog are recognized as being particularly injurious: ozone and PAN (peroxyacetyl nitrate). With PAN, injury appears on some plants as a characteristic silvering on the outer surfaces of the leaves. Ozone, on the other hand, kills cells on the upper surfaces of the leaves and appears as flecking or stippling.

Injury to plants from photochemical air pollution was noted in the mid-1940s when losses occurred to spinach and other sensitive plants in California. Since then, crop damage has occurred in many parts of the United States, particularly California, New Jersey, and Florida, and in many other countries. The damage to agriculture in the United States in 1966 was estimated at 500 million dollars. In California alone, the damage to agricultural crops during 1969 was estimated at 44.5 million dollars.

PAN

PAN is also taken into leaves through stomata and causes injury at concentrations as low as 0.01 to 0.02 ppm. In large urban areas, concentrations of 0.02 to 0.03 ppm are common. Once inside leaves, PAN attacks preferentially the spongy parenchyma cells, which collapse and are replaced by air pockets that give the leaf a glazed or silvery appearance. The symptoms on broad-leaved plants appear on the lower leaf surface, while monocot leaves show symptoms on both sides. Young leaves and tissues are more sensitive to PAN, and periodic exposures of leaves to PAN often cause "banding" and in some plants even margin "pinching" of leaves because of discolouration and death of the most sensitive affected cells, respectively.

Chlorine damage is often found near swimming pools and sewage disposal plants where chlorine is used as a disinfectant. Its damage to leaves is similar to that caused by SO_2. It is very difficult for the untrained eye to determine the cause of plant injury, since not only do the various air pollutants cause similar-appearing damage, but natural agents such as drought, frost, mineral deficiencies, and disease produce symptoms very similar to those caused by air pollutants.

Since different plant species differ greatly in their sensitivity to air pollutants, programmes for the breeding and dissemination of pollution resistant cultivars of crop plants and species of street trees are now under way. Unfortuately, this methods of control would result in loss of the use of several of our important and favourite plants.

REFERENCES

Heck, W.W., discussion to O.C. Taylor 1968. Effects of Oxidant Air Pollutants, *J. Occuptional Med.*, **10**, 485-499.

Heck, W.W., J.A. Dunning, and I.J. Hindawi. 1966. Ozone: Nonlinear Relation of Dose and Injury in Plants *Science* **151**, 577-578.

Menser, H.A. and H.E. Heggestand. 1966. Ozone and Sulphur Dioxide Synergism: Injury to Tobacco Plants *Science* **153**, 424-425.

National Research Council of Canada. "Effects of Sulphur Dioxide on Vegetation," NRC 815, Ottawa, Canada, 1939.

10

Acid Rain

Over the past decade, scientists have amassed considerable evidence that air pollutants from the combustion of fossil fuels, both oil and coal, and the smelting of metallic ores are undermining sensitive forests and soils. Damage to trees from gaseous sulphur dioxide and ozone is well documented. Recently, acid deposition, more commonly called acid rain, has emerged as a growing threat to forests in sensitive regions. Acid deposition refers to sulphur and nitrogen oxides that are chemically transformed in the atmosphere and fall to earth as acids in rain, snow, or fog, or as dry acid-forming particles. Acid deposition is known to have killed fish and plants in hundreds of lakes in Scandinavia and eastern North America. However, its links to forest damage remain circumstantial.

Temperate forests have a long history of stress and acidification. Since the end of the last continental glaciation 10,000-15,000 years ago, soils have slowly formed from the sterile layers of gravel, sand, and silt left behind by the retreating ice. Pioneering plants, animals, and microorganisms aided this soil development, helping form an intricate cycle of nutrient uptake and release. Death and decomposition of these inhabitant, generated acids in the soil. Where acids developed faster than the natural processes that could neutralize them, the soils gradually acidified. This process continues even today. Centuries of human use and abuse of forest ecosystems have added to this natural acidification. Many temperate forests in Europe and North America are now recovering from decades of intense burning, grazing, and timber cutting.

Air pollutants and acids generated by industrial activities are now entering forests at an unprecedented scale and rate, greatly adding to these stresses carried over from the past. Many forests in Europe and North America now receive as much as 30 times more acidity than they would if rain and snow were falling through a 'pure' atmosphere. Ozone levels in many rural areas of Europe and North America are now in the range known to damage trees. Despite air quality improvements made during the seventies, the average concentration of sulphur dioxide in many areas is high enough to reduce the growth of the trees.

Needles and leaves yellow and drop prematurely from branches, tree crowns progressively become thin, and, ultimately, trees die. Even trees that show no visible sign of damage may be declining in growth and productivity. Moreover, the tendency of the acid rain to leach nutrients from sensitive soils may undermine the health and productivity of forests long into the future. Taken together, these direct and indirect effects threaten not only future wood supplies but the integrity of whole ecosystems on which society depends.

Acid Rain

Normal, unpolluted rain would contain almost pure water (H_2O) in which some carbon dioxide (CO_2), some ammonia (NH_3) originating from organic matter and existing in water as NH_4^+, and varying but small amounts of cations (Ca^{++}, Mg^{++}, K^+, and Na^+) and anions (Cl^-, SO_4^-) would be dissolved. Although the pH of pure water is neutral 7.0, the pH of normal, "unpolluted" rain is usually 5.6; in other words, it is already acidic. Such rain, however, is considered "normal," and only when the pH of rain or snow is below 5.6 it is considered acidic ("acid rain").

Acid rain is the result of human activities, primarily the combustion of fossil fuels (oil, coal, natural gas) and the smelting of sulphide ores. These activities release in the atmosphere large quantities of sulphur and nitrogen oxides, which when in contact with atmospheric moisture are converted into two of the strongest acids known (sulphuric and nitric) and fall to the ground in rain and snow. The pH of rain and snow over large regions of the world ranges from 4.0 to 4.5, which is from 5 to 30 times more acid than the lowest pH (5.6) expected for unpolluted areas. The lowest rain pH's reported so far (2.4 in Scotland, 1.5 in West Virginia and 1.7 in Los Angeles) are more acidic than vinegar (pH 3.0) and of lemon juice (pH 2.2). It is estimated that about 70 per cent of the acid in acid rain is sulphuric acid, with nitric acid contributing about 30 per cent. In addition to sulphur contained in the acids carried in the rain, it is believed that an approximately equal amount of sulphur reaches all surfaces through dry deposition of particulate sulphur. In humid or wet weather, this sulphur too is oxidized to sulphuric acid.

Acid rain exerts a variety of influences by greatly increasing the solubility of all kinds of molecules and by directly (through the low pH and the toxicity of the—SO^{-3}_{4} and NO_3^{-} ions) or indirectly (through the dissolved molecules) affecting many forms of life. The adverse effects of acid rain on the microorganisms, plants, and fishes of rivers and lakes have been well documented. The effects of acid rain on crop plants have been more difficult to document. Experiments in which acidic rain (pH 3.0) was applied to plants showed that under some conditions, treated leaves developed pits, spots, and curling and that treated plants, with or without symptoms, sometimes showed reductions in dry weight. Also, more seeds of some plant species germinated when the soil in which they were planted received acid rain than when it did not, while the opposite was observed for other species. Experiments conducted to determine the effect of acid rain on the initiation and development of plant diseases have shown that in some diseases, such as *Cronartium fusiforme* rust of oak, only 14 per cent as many telia formed under acidic (pH 3.0) rain treatment than under a pH 6.0 rain treatment and that beans treated with acidic rain (pH 3.2) had only 34 per cent as many nematode egg masses than they did under a pH 6.0 rain treatment. On the other hand, a bacterial disease (halo blight) and the rust disease of bean were sometimes more severe and others milder with the acidic rain than with the pH 6.0 rain. In general, although some evidence exists that acid rain causes variable amounts of damage to at least some plants, consistent quantitative data are still insufficient to determine the extent of such damage on various crops in the areas where they occur.

The Chemistry of Acid Rain

Acid rain is a direct consequence of the atmosphere's self-cleansing nature. The tiny droplets of water that make up clouds continuously capture suspended particles and soluble trace gases. When precipitation coalesces from cloud water, it washes the impurities out of the atmosphere. All trace gases can not be removed by precipitation. However, sulphur dioxide (SO_2) and oxides of nitrogen emitted into the atmosphere are chemically converted into forms that are readily incorporated into cloud droplets: sulphuric and nitric acids.

The processes that convert the gases into acid and wash them from the atmosphere began operating long before human beings started to burn large quantities of fossil fuels; sulphur and nitrogen compounds are also released by natural processes such as volcanism and the activity of soil bacteria. But human economic activity has made the reactions vastly more important. They are triggered by sunlight and depend on the atmosphere's abundant supply of oxygen and water.

The reaction cycle occurs in the troposphere, the lowest

10 or 12 kilometres of the atmosphere. It begins as a photon of sunlight strikes a molecule of ozone (O_3) which may have mixed downward from the ozone layer in the stratosphere or may have been formed in the troposphere by the action of nitrogen, and carbon-containing pollutants. The result is a molecule of oxygen ($\dot{O}_2$) and a lone, highly reactive oxygen atom, which then combines with a water molecule (H_2O) to form two hydroxyl radicals ($H\dot{O}$). This scarce but active species transforms nitrogen dioxide (NO_2) into nitric acid (HNO_3) and initiates the reactions that transform sulphur dioxide into sulphuric acid (H_2SO_4).

The concentration of the hydroxyl radical in the atmosphere is less than one part per trillion, but it is practically inexhaustible: several of the oxidation processes it triggers end up by regenerating it. For example, one byproduct of the initial oxidation of sulphur dioxide is the hydroperoxyl radical ($H\dot{O}_3$) which reacts with nitric oxide (NO) to produce nitrogen dioxide and a new hydroxyl radical. In effect each hydroxyl radical can oxidize thousands of sulphur-containing molecules. As a result only the amount of pollutant in the air determines how much acid is ultimately produced.

Fig. 10.1 illustrates the formation of acid rain from sulphur dioxide and oxides of nitrogen given off by industry and vehicles.

The sulphuric and nitric acids formed from gaseous pollutants can easily make their way into clouds. (Some sulphuric acid is also formed directly in cloud droplets, from dissolved sulphur dioxide and hydrogen peroxide). Nitric acid gas readily dissolves in existing cloud droplets. Sulphuric acid formed through gas-phase reactions condenses to form microscopic droplets, from roughly 1 to 2 micrometres (10^{-6} m) diameter. Some of these sulphate particles settle to the ground in a process known as dry deposition. (Dry deposition also refers to the capture of sulphur dioxide gas by vegetation). Most of them, however, are incorporated in clouds. Moisture readily condenses on an existing surface—a condensation

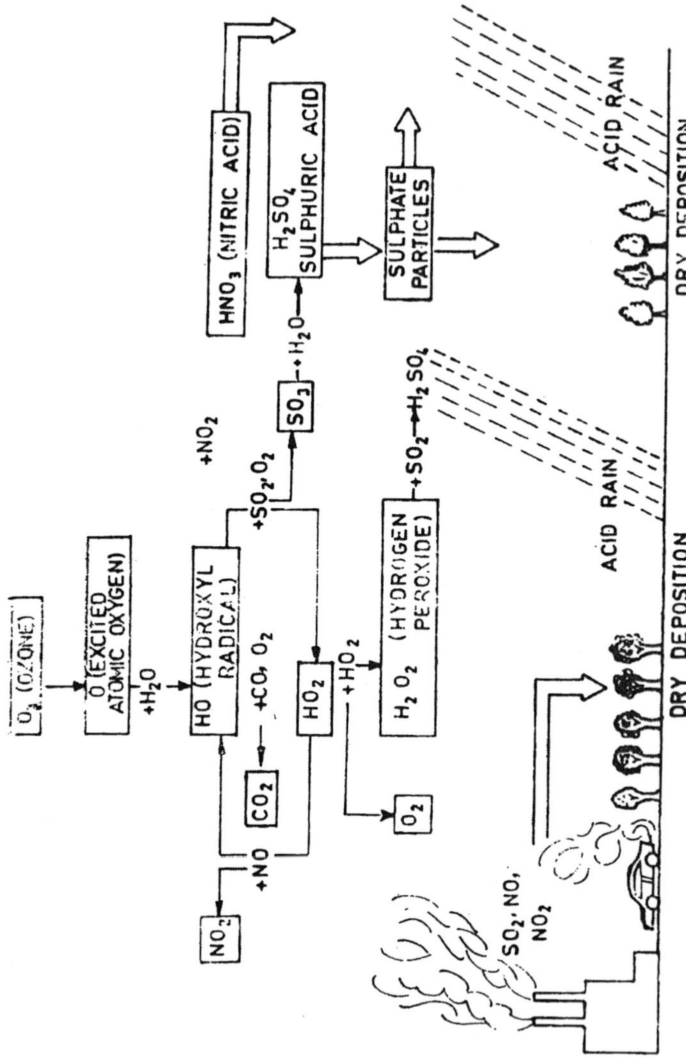
O_3 (OZONE)
O (EXCITED ATOMIC OXYGEN)
$+H_2O$
HO (HYDROXYL RADICAL)
$+CO, O_2$
CO_2
HO_2
$+NO$
NO_2
$+HO_2$
O_2
H_2O_2 (HYDROGEN PEROXIDE)
$+SO_2, O_2$
SO_3
$+H_2O$
$+NO_2$
HNO_3 (NITRIC ACID)
H_2SO_4 SULPHURIC ACID
SULPHATE PARTICLES
$+SO_2$
H_2SO_4
ACID RAIN
ACID RAIN
SO_2, NO, NO_2
DRY DEPOSITION
DRY DEPOSITION

nucleus—and sulphate particles are ideal condensation nuclei. They grow into cloud droplets containing dilute sulphuric acid.

The sulphuric and nitric acids in cloud droplets can give them an extremely low pH. Water collected near the base of clouds in the eastern U.S. during the summer typically has a pH of about 3.6, but values as low as 2.6 have been recorded. In some parts of the world the pH of fog has fallen as low as 2 which is the same as that of the acid-lemon juice.

The acid rain may fall hundreds of miles from the pollution source. Wherever it lands, it undergoes a new round of physical and chemical changes, which can reduce the acidity and alter the chemical characteristics of the water that eventually reaches lakes and streams. Alkaline soils, such as soils rich in limestone, can neutralize the acid directly. In the slightly acidic soils typical of the evergreen forests exposed to acid rain in the U.S., Canada and Europe two other processes can counter the effects of acid deposition. The acid can be immobilized ast he soil or vegetation retains sulphate and nitrate ions (from sulphuric and nitric acids respectively). In can also be buffered through a process that is known as cation (positive ion) exchange.

In cation exchange the ions of calcium, magnesium and other metals found in many soils take the place of the acid's hydrogen ions. The source of the metal ions is rock weathering: the dissolution of minerals by precipitation and groundwater containing dissolved carbon dioxide, which releases the positive metal ions into the soil together with anions, or negative ions of bicarbonate (HCO_3^-). Then, when sulphuric acid is added to the soil, the sulphate (SO_4^{2-}) of the acid can displace the calcium or magnesium ions. As the sulphate solution washes the metal cations from the soil, the hydrogen ions responsible for the acidity are left behind.

Acid Rain and Forests

The role of acid rain and other forms of air pollution is

under intensive investigation. In spite of the dimensions of the forest damage, however, a firm link has not been established. One can get some idea of the difficulties by contrasting the forest decline with clear-cut cases of forest poisoning by air pollutants. Smelters and chemical plants that emit sulphur dioxide, oxides of nitrogen or fluoride compounds are often girdled by dead timber. In such cases there is a clear correlation between tree damage, a specific pollution source and a threshold concentration of the pollutant. The forests that are now dying, in contrast, are far from any source and are exposed to pollutants in concentrations well below the levels previously reported to injure trees. If air pollution and specifically acid rain, plays a part in forest decline, it probably does so less as a lethal agent than as a stress.

Many stresses, both biotic and abiotic, combine to affect the vigour of a forest. The trees' genetic endowment or age can be a source of stress: a stand may be genetically weak or senescent. Other stresses may take such forms as diseases, insects, parasitic fungi, a shortage of light, water or essential nutrients and sporadic injury from events such as floods, high winds and ice storms. Stresses easily withstood in isolation can combine with debilitating or fatal effects. A fatal sequence of stresses may begin with a "predisposing" stress, such as a shortage of nutrients. The tree may then be seriously weakened by an "inciting" stress, such as a severe winter. It is then defenseless against a final, "contributing" stress—the actual cause of death—such as disease or insect attack.

Acid and other pollutants could add to the high level of abiotic stresses, including thin soil, low temperatures and desiccating winds, present in a high-elevation forest. That is, the pollutants might handicap the trees with one more predisposing stress as they face subsequent stresses. But what is the nature of the added stress?

Investigators, most of them in Europe, have put forward a

number of hypothetical mechanisms, many of which would ultimately lead to nutrinct deficiency in the tree.

Sulphates and nitrates raining down as acids have drastically different effects on different forests and even on different tree species in the same forest. Incoming acids affect interactions between the soil and living biomass of an ecosystem in complex and varying ways. Soil structure and composition, vegetation type, climate, and elevation are only some of the natural determining variables. Yet, research over the last decade has uncovered some common effects of acidity that point to several pathways by which acid deposition can threaten forests.

Trees derive their nutrition primarily from elements such as calcium, magnesium, and potassium that are weathered from minerals in the soil. Acid deposition adds hydrogen ions to the soil, which displace these important nutrients from their sites bound to soil particles. Soils with a pH of 5 or more are seldom in danger since they have plentiful calcium carbonate (the constituents of lime) or silicates (which have abundant calcium, potassium, and/or magnesium) that effectively neutralize the acid ions. Yet soils at lower pH levels have fewer of these buffering agents. Competition from incoming acids causes the leaching of calcium and magnesium from the soil. Large areas of the southeastern United States, New England mountain ranges, eastern Canada, and extensive areas in Scandinavia, for example, are underlain by slightly acidic, poorly buffered soils that are especially susceptible to the effect. Although soil changes generally take place over a long period of time, soil studies in Sweden suggest that substantial leaching of nutrients from sensitive soils can occur in just a decade.

Sulphates and nitrates, the other key constituents of acid deposition along with hydrogen ions, can initially have a fertilizing effect on many soils and for a time may actually boost tree growth. Yet this enhanced growth is short-lived, for eventually these "fertilizer" supplies will exceed the forests' capacity to use them. Sulphate saturation usually precedes nitrate saturation, but excess quantities of either or both simply pass through the soil carrying vital nutrients with them. With forest productivity closely tied to nutrient availability, this leaching of soils by hydrogen, sulphate, or nitrate ions eventually reduces forest growth.

Research has also shown that heavy metals—either mobilized in the soil or introduced from the atmosphere—may be involved in the forest damage. Bernhard Ulrich, a soil scientist who has studied damaged beech and spruce forests in the Solling Plateau of West Germany for nearly two decades, has proposed that as soils become increasingly acidic, aluminium, which is normally harmlessly bound in soil minerals, becomes soluble a nd toxic. The free aluminium attacks the tree's root system, making the tree less able to take up moisture and nutrients and to protect itself from insect attacks and droughts.

Fig. 10.2 illustrates the mechanism by which acid rain and ozone together could lead to nutrient deficiency in a coniferous tree, according to a currently favoured mechanism for their possible role in forest decline.

Control

The most direct way of controlling the pollutants that cause acid rain would be to burn less fossil fuel for transportation and energy generation. Rapid mass transport system and fuel-efficient cars can reduce oil consumption in the transportation sector, but energy generation is less manageable. In spite of worthy strategies for conserving energy, consumption is likely to increase in the long run, and current alternatives to fossil-fueled power plants do not look promising. Hydroelectric power is limited by a variety of problems, including scarcity of appropriate sites, resettlement of people, deforestation etc nuclear power is beset by economic problems and lack of public confidence in its safety.

The key to controlling acid rain, then, must be the reduction of emissions from fossil-fueled power plants, particular coal-

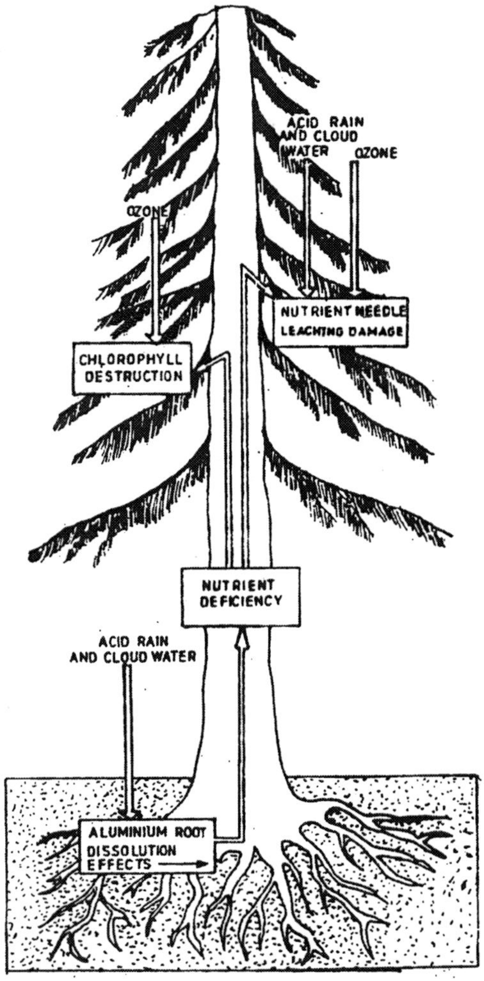

Fig. 10.2 Acid Rain and Ozone Damage to Trees

burning power-plants. The approach that has already led to reductions in sulphur emissions in the U.S., West Germany and Japan combines the use of coal that is naturally low in sulphur, or has been washed to remove sulphur and other contaminants, with flue-gas desulphurization (FGD). In FGD wet limestone is sprayed into the plant's hot exhaust gases, where it removes as much as 90 per cent of the sulphur dioxide. The sulphur-containing waste can be difficult to dispose of, however, and FGD reduces the efficiency of a power plant, causing it to consume several percent more coal for a given output. Furthermore, the process does nothing to reduce nitrogen oxide emissions.

The new power-plant technologies developed USA., jointly by the Government and industry under the ' Clean Coal Demonstration Programme, enacted in 1984, offer a more comprehen sive solution. Three clean-coal technologies are already being demonstrated in full-size plants. In the system known as atmospheric fluidized-bed combustion, a turbulent bed of pulverized coal and limestone is suspended by an upward blast of air; the combustion region is threaded with boiler tubes, which supply steam to the plant's turbines. The turbulent mixing of the coal and the air allows combustionto take place at a lower and more even temperature than it does in a conventional boiler, which reduces the formation of oxides of nitrogen, Meanwhile the limestone efficiently captures the sulphur dioxide. In a related technology known as pressurized fluidized-bed combustion the coal is burned in compressed air, which improves the plant's efficiency as well.

In the third technology, gasification/combined-cycle, coal is reacted with steam and air at high temperatures to produce a gas consisting mainly of hydrogen and carbon monoxide. The gas can then be burned, spinning a turbine; waste heat in the gas turbine's exhaust serves for generating steam, which drives a steam turbine to yield additional electricity. A gasification/

combined-cycle plant operates much more efficiently than a conventional plant and gives off considerably less sulphur dioxide and nitrogen oxides.

REFERENCES

Abrahamsen, G. 1984. 'Effects of acidic deposition on forest soil and vegetation.' *Philosophical Transactions of the Royal Society of London* 305, 369-382.

Balzhiser R.E. and Kurt E., Yeager; 1958 "Coal-fired Power Plants for the future," *Scientific American* 258, 92-96.

Blank, L.W. 1985. 'A New Type of Forest Decline in Germany.' *Nature* 314, 311-314.

Brady, G.L. and Selle, J.C. 1985. 'Acid Rain—The international response'. *International Journal of Environmental Studies* 24, 217-230.

Buttler, J.D. 1979. *Air Pollution Chemistry*, Academic Press, New York.

Park, Chris. C. 1987. *Acid Rain-Rhetoric and Reality*. Methuen & Co., New York.

Woodman, M. James E.B. [Cowling. 'Airborne chemicals and forest health ' *Environmental Science and Technology*, 21, 120-126.

11

Air Pollution and Human Health

The air pollution and human health studies involves a large number of variables. Consequently it becomes difficult to prove that air pollution has a clearly demonstrable effect on human health at 'tolerable' urban concentrations of air pollutants. It is well established that at the elevated levels air pollutants can cause severe changes in human physiology and a quick death. It is also clear that air pollution has adverse effects on those who already have respiratory problems. However, it is difficult to show that air pollution is the basic cause of the disease. In this chapter some adverse effects of ambient levels of air pollutants on human health are discussed.

Numerous studies of the effects of gaseous pollutants and particulates on animals have been carried out. Most of such studies have been done at very high concentrations and thus have no direct significance for humans since the concentrations of pollutants in human habitats are lower. Furthermore, there is a wide variety of response by different animal species to the same concentration of pollutant. Thus, any extrapolation of these effects is a complicated exercise. For example, the guinea pig, expected to be most susceptible to sulphuric acid aerosols, with stand concentrations which would be intolerable to man.

Although one cannot yet implicate most ambient air pollution level as a source of illness, it is generally acceptable that increasingly smaller dose can be shown to have an effect.

Pollutants may attack the lungs through five main cites of action or mechanisms:

1. in the respiratory system, which reacts by the initiation of a constructive reaction of a bronchi reflex;
2. in the blood vessels of the bronchus and its branches, which try to reduce the absorption of harmful substances through the bronchial mucosa;
3. in the blood vessels of the lungs, where they react by decreasing absorption from the alveolar (pulmonary) capillaries;
4. in the heart and large blood vessels taking part in the transport of toxic substances;
5. by penetration into organs, tissues or cells and by affecting metabolic processes.

In order to understand this let us first give a brief description of the anatomy of the human respiratory system and the mechanisms by which it cleans itself. This is followed by the effects of air pollutants on human health as determined by epidemiological studies. Some terms used in this connection are given in Table 11.1.

The Respiratory System

The human respiratory system functions to take in oxygen needed by the metabolic processes and to remove carbon dioxide produced in the body. The exchange of these two gases takes place in the lungs. Blood acts as the carrier to transport oxygen to the tissue and to bring carbon dioxide to the lungs.

TABLE 11.1 **Some Important Terms**

Alveoli (pl.) Alveolus (sing.)	Small, sac-like dilations at the innermost end of the airway, through whose walls gaseous exchange takes place.
Bronchiole	One of the finer subdivisions of the bronchial tree.
Bronchitis, chronic	A long standing inflammation of the bronchi, manifested by cough and the production of sputum.
Bronchi (pl.) Bronchus (sing.)	One of the larger air passages in the lung.
Cilia	Small, hairlike process attached to the free surface of a cell, capable of rhythmic movement.
Emphysema	A condition in which there is overdistention of air spaces and resultant destruction of alveoli and loss of functioning lung tissue.
Epidemiology	A science dealing with the factors involved in the distribution and frequency of a disease process in a population.
Morbidity	The occurrence of a disease state.
Mortality	The ratio of the total number of deaths to the total population, or the ratio of the number of deaths from a given disease to the total number of people having that disease.
Synergism	A situation in which the combined action of two or more agents acting together is greater than the sum of the action of these agents acting separately, also called potentiation.
In vivo	Within the living body.
In vitro	Within a test tube or artificial environment.

We breathe about 15-18 times per minute at rest and take in about 0.5 litres of air per breath. Most of this air is mixed in the lungs with air that is already there. However a portion remains in the "dead space", which is the name given to the large air spaces in the respiratory system where no gaseous exchange takes place.

Most of the oxygen and carbon dioxide transported by the blood are not dissolved in the plasma but instead combine with the haemoglobin. Oxyhaemoglobin is the name given to haemoglobin to which oxygen is attached. Haemoglobin in which carbon monoxide has become attached is known as carboxyhaemoglobin, COHb. Since the affinity of CO for haemoglobin is about 200 times more than that of O_2, it is easy to understand why low levels of CO can still produce significant amounts of carboxy haemoglobin.

The respiratory system is shown in Fig. 11.1. The

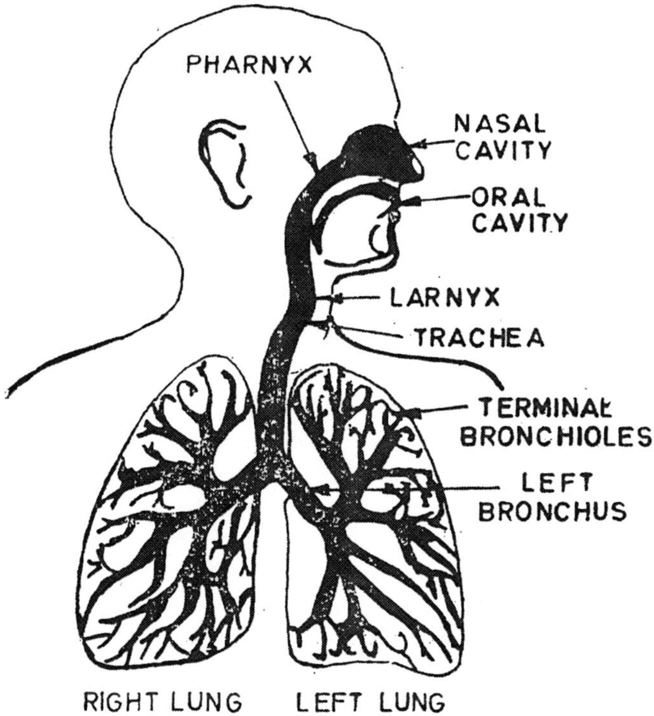

Fig. 11.1 The major anatomical features of the human respiratory system. (The diagram shows the major divisions of the human respiratory tract into nasopharyngeal, tracheobronchial, and pulmonary compartments).

nasopharyngeal section, the tracheobronchial system, and the pulmonary structure are the three main sections. The nasal passages lead through the nasopharyngeal structure (throat) to the trachea and bronchi. The bronchi break up into smaller bronchioles which then terminate in the alveolar sacs. The exchange of oxygen and carbon dioxide between the air and the blood, occurs in the alveolar sacs. Presence of a large number of these tiny sacs provide a very large surface area for this exchange.

The airways narrow as the air goes from the trachea to the alveoli. However, the increase in cross-sectional area more than compensates, and the velocity of the air slows as it penetrates into the lungs. The bronchial and alveolar structure of the lungs, are shown in Fig. 11.2.

Another very important part of this structure is the cilia shown in Fig. 11.3. The nasopharyngeal and the tracheobronchial tubes have a layer of cells which have many small hair-like projections on them called the cilia. These beat in a rhythmic movement to produce a wave motion upward which carries mucus and entrapped particulates, germs, and dissolved gases away from the lungs. This material is then either swallowed or coughed up when the mucus is carried out of the trachea. The pulmonary surface is not ciliated and has a moist layer of cells covered by a material which prevents the collapse of the alveoli at the end of respiration. The inhaled pollutants, both gaseous and particulate thus have several opportunities to be removed before entering the alveoli. Gases can be absorbed along the pharynx or bronchioles and particulate can be deposited in the mucus layer and then removed through the action of the cilia.

The removal mechanism or clearance processes are important in considering the effect of air pollution on the respiratory system. The material removed may be taken up by the blood stream, by the gastrointestinal tract via ciliary clearance of the bronchioles, or by the lymph system.

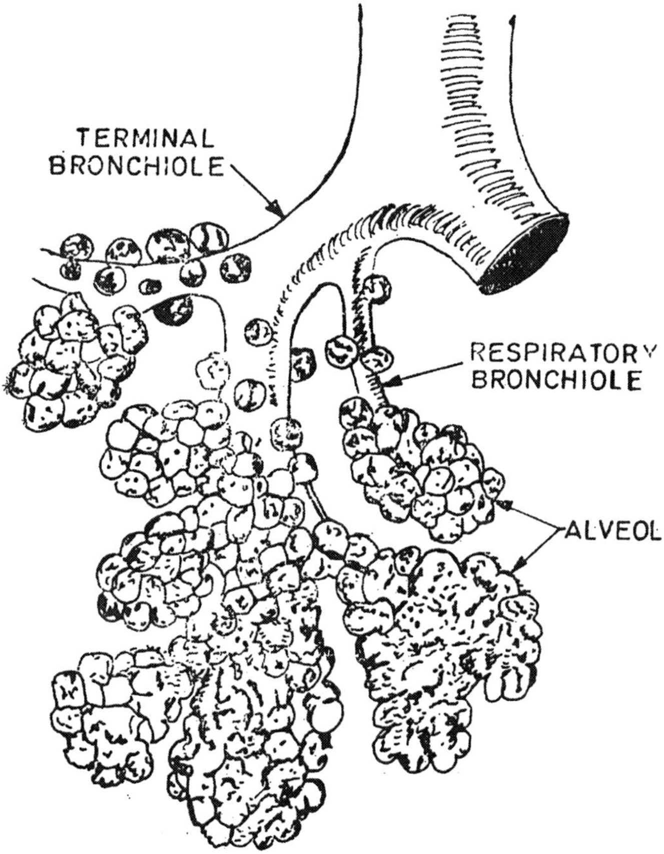

Fig. 11.2. The terminal bronchial and alveolar structure of the human lung. (The diagram shows the pulmonary structure of the respiratory tract).

The cilia action is quite rapid, Experimental results indicate that clearance occurs in two phases, one of few hours and one of many hours. The respiratory system can clean itself to a large extent. Even some of the deposits on the alveolar surface can eventually reach the ciliary mucus transport system (via the action of the microphage cells) or the blood stream, by penetrating through the alveolar wall.

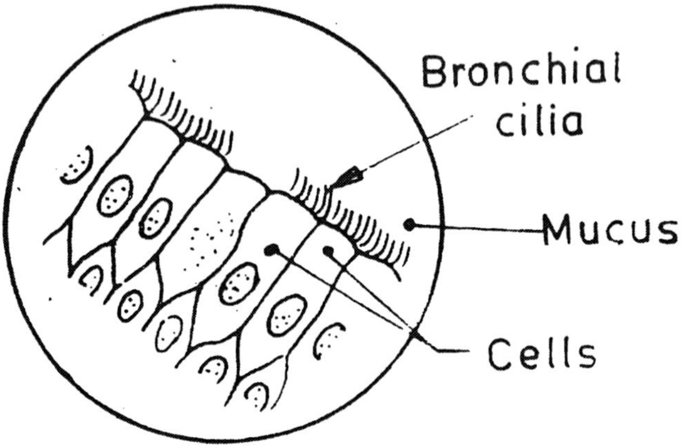

Fig. 11.3

Obviously, it is the continuous exposure to high concentrations that overloads the clearance mechanisms and results in high equilibrium concentrations of deposited material in the respiratory system.

From this discussion it is clear that most of the effects of air pollutants on health are related to respiratory disease, i.e., bronchitis, lung cancer, etc. Animal studies of individual pollutants have also shown that at high concentrations specific effects can occur; for example, the cilia may stop beating and thus cut off the main cleaning mechanism of the system. In short, there are definite relationships between common air pollutants and physiological effects on the respiratory system.

Specific Diseases

Lung Cancer

According to Lave and Seskin (1970) there is definite evidence of an "urban" effect on lung cancer. A mong smokers the likelihood of dying from lung cancer ranges from 26 per cent to 123 per cent greater if one lives in an urban area. For nonsmokers the range is wider and the urban effect appears to be even greater. The obvious question is: is this increased incidence of urban lung cancer caused by air pollution or is it at all related to air pollution?

Lave and Seskin (1970) argue that no reasonable man can object to the evidence of a substantial quantitative association between air pollution and several diseases, not just lung cancer. They further argue that the possible effects for city dwellers of greater smoking, occupational exposure, less exercise, additional tension and stress are not an explanation. It seems reasonable to conclude thot while the above factors may have some effect, air pollution must also play an important role.

Bronchitis

The reviews by Goldsmith (1968), Lave and Seskin (1970), and the British Royal College of Physicians survey indicate that exposure to air pollution in the form of suspended particulates and sulphur oxides is a causal factor in promoting and aggravating bronchitis. Smoking habits and population density are also important variables found in epidemiological studies.

Other Cancer

There is evidence of an association between stomach cancer and air pollution. Remember that the cilia clearing action removes particulate from the bronchioles after which they are swallowed. Several chemicals which are carcinogenic (cancer-causing) have been found in the atmosphere. The chemical of most interest as a cancer-causing agent is benzo (a) pyrene which is one of the polynuclear aromatic hydrocarbons. This family of hydrocarbons may exist in the atmosphere as free molecules or can be adsorbed on carrier particles. The size of the carrier particulate affects its possibility of deposition in the lungs. In urban atmospheres there is an ample supply of particulate of about 1 micron size for deposition in the lung.

Some Specific Effects

Tolerance

Some pollutants seem to have the ability to initiate a tolerance response in laboratory animals. In particular, rats and mice exposed to ozone are found to develop a protection to second exposures. In fact, ozone exposure can provide protection to later exposure of NO_2, H_2O_2, H_2S, and other gases. Experiments with mice exposed to NO_2 also indicate reduced effects upon subsequent exposure.

Cilia Effects

Since the cilia are the primary mechanism by which the bronchiole tubes are cleared of particulate any effects of gaseous pollutants in reducing their effectiveness is of interest. At SO_2 concentrations of 12 ppm (far higher than ambient level) ciliary cessation has been shown to occur in about 4 to 6 minutes in rats. After exposure the cilia regained their mobility. Cigarette smoking can also stop the cilia action.

Increased Infection Susceptibility

Air pollutants are known to increase the susceptibility to bacterial infection. For example, NO_2 exposure at 0.5 ppm for several months caused increases in mortality of mice exposed to *Klebsiella pneumoniae* (Ehrlich and Henry, 1968). Other studies, at higher NO_2 levels, on squirrels, monkeys and hamsters showed similar effects. Ozone has also shown an ability to increase susceptibility to bacterial infection. Mice exposed to streptococcal aerosol after a 3-hour exposure to ozone from 0.08 ppm to 0.52 ppm showed an increase in mortality. Mice and hamsters exposed to higher ozone levels and then exposed to *K. pneumoniae* also exhibited increased mortality. Thus it is certainly possible that air pollutants can increase human susceptibility to respiratory disease and mortality.

Specific Pollutants

Carbon Monoxide

Carbon monoxide is an odourless, tasteless, and colourless gas. It is produced as a product of incomplete combustion. Upon entering the respiratorv system it combines in the lung with the haemoglobin in the bloodstream to form carboxyhaemoglobin, COHb. This reduces the ability of the haemoglobin to carry oxygen to the body tissues. The affinity of CO is some 200 times that of O_2 for attaching itself to the haemoglobin so that low levels of CO can still result in high levels of COHb.

After exposure the CO is slowly released from the blood with a clearance half-life of 3 to 4 hours. Cigarette smokers commonly have COHb levels of about 5 per cent with heavy smokers reaching values of 10 per cent. This is due to the high CO concentration, about 400 ppm, in cigarette smoke.

At COHb levels of 2 to 5 per cent effects are found to the central nervous system such as impairment of time interval discrimination and visual acuity. At levels greater than 5 per cent there are cardiac and pulmonary function changes. There is evidence that CO concentrations encountered in the Los Angeles area (10 to 15 ppm ambient Los Angeles *basin* 24-hour *average* values) are associated with excess mortality (Hexter and Goldsmith, 1971). Evidence has also been prcsented that a relationship exists between ambient CO levels and myocardial infarction (heart attacks) by Cohen et al. (1969). There is ample evidence that ambient air concentrations of CO in major traffic areas are high enough to produce significant effects on health. Some properties of pollutants and their effects are listed in Table 11.2.

Ozone

Ozone is extremely toxic. Rats exposed to 1 ppm for an 8-hour day develop bronchitis, fibrosis (formation of fibrous

TABLE 11.2 **Properties and Effects of Pollutants**

Ozone	Colourless	
	Odour threshold	0.02-0.05 ppm
	Irritation of nose and throat	0.05 ppm
	Headache (in 30 minutes)	1 ppm (1962 μgm^{-3})
	Industrial limit (WHO 8-h day)	100-200 μgm^{-3}
NO_2	Brown-red colour	
	Odour threshold	0.12-0.22 ppm
	Increased respiratory illness over 6-month test period	0.06-0.109 ppm
	U.S. industrial limit (8-h day)	0.5 ppm
SO_2	Colourless	
	Taste threshold	0.3 ppm
	Odour threshold	0.5 ppm
	Increased mortality	0.25 ppm (24 h measurement) with 750 $\mu g\ m^{-3}$ smoke
	Possible increased mortality	0.19 ppm (24 h measurement) with "low" particulate levels for "few" days
	Increased respiratory illness in children	0.046 ppm with smoke at 100 $\mu g\ m^{-3}$ for long term exposure
	Throat irritation	8-12 ppm
	Eye irritation	10 ppm
	Immediate coughing	20 ppm
	Industrial limit (WHO, 8-h day)	10-13 mg m^{-3}
CO	Colourless, tasteless odourless (faint odour at very high concentrations).	
	Central nervous system effect	15 ppm for 10 h.
	Industrial limit (WHO 8-h day	50 ppm

Source: Adapted from Stern, A.C. *Air Pollution*—Vol. I 2nd edn. Academic Press 1968.

tissue), and bronchiolitis. A concentration of 1.25 ppm for 1 hour causes an increase in the residual lung volume (reduced expiration) and a decrease in the breathing capacity. Higher levels cause pulmonary oedema (accumulation of fluid in the lungs), haemorrhaging, and impairment of gas exchange through the alveolar membrane. The maximum allowable concentration for occupational exposure for an 8-hour work day has been set at 0.1 ppm.

A curious characteristic of ozone toxicity is that tolerance develops in laboratory animals from exposure to subinjurious concentrations. The tolerance develops quickly, within 24 hours. It may persist for 4 to 6 weeks in rats and 100 days in mice, as shown by their survival from exposure to concentrations that would ordinarily be lethal. After tolerance develops, the pulmonary oedema and haemorrhaging that usually follow strong dose exposures are absent. Moreover, cross-tolerance can be demonstrated between ozone and the lethal gases phosgene and nitrogen oxides.

Ozone increases susceptibility to infection. Experiments with mice showed that ozone increases the survival of bacteria inside the lungs. In rabbits, cells washed from the lungs after exposure to ozone showed that there was a reduction in the ability of the alveolar macrophages to engulf bacteria. Also, laboratory animals that have been subjected to prolonged exposure to ozone show signs of premature aging.

The injurious effect of ozone may be partly due to its ability to oxidize lipids (fats). A deficiency of tocopherol (member of the vitamin E complex containing several tocopherols) is known to make tissues more susceptible to oxidation. Rats with a deficiency of vitamin E are more susceptible to lethal concentrations of ozone than those not deficient.

Oxides of Nitrogen

Both oxides of nitrogen, NO and NO_2, are potential health hazards. Animal mortality studies indicate that NO_2 is about four times more toxic than NO. No cases of human death from NO poisoning have been reported, at the concentrations found in the atmosphere. NO is not an irritant and is not considered a health hazard. The greatest toxic potential of NO is its ability to undergo oxidation to the more toxic NO_2.

The proven effects of NO_2 on humans and animals are confined almost entirely to the respiratory tract and occur only with NO_2 levels higher than those now prevailing in the atmosphere. A higher NO_2 dose results in the following effect sequence: nasal irritation, breathing discomfort, acute respiratory distress, pulmonary oedema. and finally death. Even the mildest effects, such as mucous membrane irritation, do not occur at the prevailing atmospheric NO_2 concentrations. The odour threshold range for NO_2 in human is reported to be 1-3 ppm. Concentration above 12 ppm resulted in eye and nasal irritation in a small number of volunteers. In these cases, the nasal irritation was more intense than that of the eyes. Concentrations of NO_2 greater than 100 ppm are lethal to most animal species, and 90 per cent of the resulting deaths are due to pulmonary oedema.

Oxides of Sulphur

Sulphur dioxide is intensely irritating to the eyes and the respiratory tract. A concentration of 0.3 to 1 ppm can be detected by the average person, and this may be by taste as well as smell. Most people will show a response from exposure to concentrations in the range of 1 to 5 ppm. Exposure to the higher concentration for 1 hour causes a pronounced choking sensation, and 10 ppm for the same period produces severe distress. Exposure can cause serious inflammation of conjunctiva of the eyes. Concentrations of 6 to 12 ppm cause immediate irritation of the nose and throat. A few minutes of exposure to 1 per cent (10,000 ppm) produces irritation, even to moist areas of the skin.

Sulphur dioxide is so irritating that at acutely dangerous concentrations, it serves as its own warning agent. However, this safety factor does not always exist at low concentrations. Moderate resistance develops from continuous exposure so that concentrations as high as 5 ppm can no longer be smelled. Concentrations of 400 to 500 ppm may be lethal and are considered to be immediately dangerous to life, while 50 to 100 ppm is regarded as the maximum permissible range for 30 to 60 minutes exposure. The accepted maximum allowable concentration for an exposure of 8 hours is 10 ppm.

Moist air and fogs probably increase the danger of sulphur dioxide greatly, partly due to the formation of sulphurous acid and sulphite ions. A further explanation may be that sulphur dioxide becomes adsorbed on finely divided aerosol particles, resulting in deposition deep in the air spaces of the lungs. The toxic end product of sulphur dioxide is probably sulphuric acid, which is a stronger irritant than sulphur dioxide. Sulphuric acid causes 4 to 20 times the effect on an equimolar basis (molecule for molecule).

Particulates

Particulate pollutants enter the human body almost exclusively through the respiratory system. Hence they effect this system almost immediately. The intensity of the effects depend on the degree of particulate penetration within the system, and the particulate toxicity.

In the upper respiratory tract, particles greater than 5.0 μ (5000 nm) in diameter are filtered from the inhaled air. Particles smaller than this size may escape the defense mechanism of the upper respiratory tract and enter the lungs. Those between 0.5 to 5.0 μ (500—5000 nm) may be deposited as far as the bronchioles, but few reach the alveoli. Such particles are usually removed within a few hours by cilia as we have already seen. Particles less than 0.5 μ (500 nm) in diameter reach the alveoli and may settle there. The removal of such particles from these areas is much less rapid and less complete than from the larger passages since alveolar membranes have no mucous covering.

Particulate matter that enters and remains in the lungs can exert a toxic effect in three different ways:

1. Particles that are themselves inert may interfere with clearance mechanisms in the respiratory tracts and prevent or slow the removal of other harmful particles.

2. The particles may carry absorbed or adsorbed gas molecules and thus enable them to reach and remain in the sensitive areas of the lungs.

4. Particulates may be intrinsically toxic and as a result, directly affect the body. Such particulates are rarely found in the air in high concentration. However, many toxic materials are present in trace amounts. Much of the concern about these substances is related to the possibility of their concentrations increasing beyond current levels.

Asbestos Dust

Asbestos dust is an air pollutant which constitutes a considerable localized hazard of the manmade environment. Asbestos is a fibrous material made up of magnesium silicate. It is used in building and in engine gaskets and brake linings. Its main value lies in its resistance to fire, heat and chemical attack. There are two main forms of asbestos, blue asbestos (crocidolite) and white asbestos (amosite, chrysotile, tremolite). Blue asbestos dust is particularly dangerous but dust from both types of asbestos can have serious effects following inhalation. These effects include lung scarring (asbestosis) and cancer of the bronchi pleura and peritoneum. Although threshold limiting values for asbestos dust in the atmosphere have been established, it seems likely that no concentration can be regarded as completely safe.

Fluorides

Fluoride particles are released into the atmosphere by brick factories, aluminium smelters and phosphate works. These particles settle on surrounding pasture and there are many recorded observations of cattle losing teeth and suffering lameness from bone malformations following ingestion of fluoride during grazing. This type of poisoning has been referred to as fluorosis. Plants may also suffer from fluoride poisoning. Gladioli of the variety 'Snow Princess' are particulary sensitive. The leaves mottle and turn brown on exposure to concentrations of hydrogen fluoride of 0.5 $\mu g\ m^{-3}$ or more. There are reports of damage to maize and citrus plants at concentrations as low as 4 $\mu g\ m^{-3}$ in the atmosphere. Since atmospheric fluoride concentrations of 1 $\mu g\ m^{-3}$ can lead to plant fluoride concentrations in excess of 30 ppm, it will be seen that chronic exposure to low levels of fluoride may have harmful effects because of fluoride accumulation. Dairy cattle may be particularly at risk.

Deliberate fluoridation of drinking water by public health authorities, with the intention of preventing tooth decay has been a matter of some controversy. Fluoride, as described above, can be very toxic but, so far, no harmful effects of fluoride have been conclusively demonstrated at the concentrations present in drinking water. Possible correlations with increased incidence of disease have been suggested, but other explanations of the statistical evidence are possible. However, the question remains as to whether the benefit of reduced tooth decay is sufficient to warrant even the remotest possibility of an increase in other, more serious illnesses.

REFERENCES

Bove, J.L., and S. Siebenberg. 1970. 'Airborne Lead and Carbon Monoxide at 45th Street, New York City'. *Science* 167, 986-987.

Chow, T.J. and J.L. Earl. 1970. 'Lead Aerosols in the Atmosphere: Increasing Concentrations'. *Science* 169, 577-580.

Cohen, S. I., M. Deane, and J.R. Goldsmith. 1969. Carbon Monoxide and Survival from Myocardial Infarction, *Arch. Environ. Health*, 19, 510-520.

Ehrlich, R. and M.C. Henry. 1968. 'Chronic Toxicity of Nitrogen Dioxide: Effects on Resistance to Bacterial Pneumonia'. *Arch. Environ. Health*, 17, 860-865.

Goldsmith, J.R., 1968. "*Air Pollution*," vol. 1, Academic Press, New York.

Hexter, A.C. and J.R. Goldsmith. 1971. 'Carbon Monoxide: Association of Community Air Pollution with Mortality'. *Science*, 172, 265-267.

Lave, L.B. and E.P. Seskin. 1970. 'Air Pollution and Human Health'. *Science 169*, 723-733.

12

Water Pollution

"The solution to pollution is dilution" is an old saying. Even today, we depend mainly on dilution for disposing of wastes in air, water, and soil, for some of the waste problems in nature are taken care of by the large amount of living space. Unfortunately, when living space in any part of the biosphere becomes limited, as, for example, by congestion or industrial expansion, dilution is no longer a satisfactory solution. Hence other means must be taken to reduce the ratio of waste to space.

Most of the wastes of civilization are drained into streams and rivers and ultimately into lakes or oceans. In many cases the waste is dumped into the same bodies of water from which drinking water is withdrawn. The effect of sewage, industrial waste, and agricultural drainage on plant and animal life in closed bodies of water is sometimes catastrophic. The accumulation of excessive plant nutrients is called *eutrophication.*

Coastal waters are also changing due to organic material from sewage outfalls. Marine algae are disappearing from sections of coastal areas. Kelp, an important source of algin used in food and medicines, is diminishing. Heavy metals and pesticide residues are found in fish in amounts above those declared to be safe for human consumption. Some species of fish-eating birds accumulate DDT in large quantities and the pesticide is suspected of being the cause of thin egg shells and the decline of certain species in some of their breeding grounds. The number of other species of commercial fish and marine birds are decreasing at an alarming rate. The specific cause is unknown in most cases.

The vastness of the earth's water had led many people to think of the rivers and eventually the oceans as the ideal receptacle for waste—a garbage dump of seemingly unlimited dimensions. But the delicate balance of many organisms in their environment is often overlooked. Extraneous materials deposited along or near the shores may remain in concentrated form for long periods of time. Some of the pollutants can disrupt the life cycle of plants and animals or readily enter into the food chain with potentially serious or even disastrous results.

The oceans, which cover 70 per cent of the earth's surface, receive an estimated 4 billion tons of dissolved matter annually. Some of it is man-made. It is essential to find out what it contains and the effect of each constituent on water, wildlife, and people.

Eutrophication

Lakes, large and small, are constantly being enriched by organic matter from decomposing plant and animal remains, providing nutrition for algae and larger aquatic plants. This fertilization, whether from natural sources or man-made, is called *eutrophication*, from the Greek meaning "well fed." This condition occurs often in nature and is not necessarily bad, for the organic material means food for fish and other aquatic life.

But too much mineral nutrient in the water can be more serious than too little. Many people have noticed the reddish or greenish colour of lakes, bays, or coastal areas in summer or fall caused by enormous numbers of algae and other phytoplankton. Their presence is often followed by the death of fish and other aquatic life, sometimes on a massive scale. Such occurrences are the result of microscopic life having become overabundant. When the algae and other plankton die, decomposition of the superabundance of organic material causes the oxygen in the water to become suddenly depleted. Water cannot absorb replacement oxygen fast enough to take care of the needs of the living organisms. Thus, they suffocate. Also, toxins are sometimes produced.

The "red tide" that is frequently seen along the coasts of California, the Gulf of Mexico, and India is caused by the buildup of microscopic ***dinoflagellates***. These are single-celled swimming organisms that have some of the attributes of both animals and plants. They impart a red or dirty brown colouration to the water and at night create a spectacular display of luminescence in the froth of waves. Their most injurious effect is the production of a lethal poison called ***saxitoxin***, which accumulates in the bodies of clams and mussels. People who eat them are liable to be stricken with "paralytic shellfish poisoning," a dangerous form of food poisoning having a 21 per cent mortality rate. The red tide along the coast of India is thought to be caused by an overabundance of mineral nutrients deposited by the heavily mineral-loaded run-off of flood water during the monsoon rains. Along the California coast, it is suspected that the nutrients from sewage outfalls contribute to excessive growth of the organisms when suitably high temperatures and other favourable conditions prevail.

In contrast to many of the land areas where the scarcity of water is the chief limitation to the growth of vegetation, the oceans are afflicted with a lack of nutrients. This is the most critical factor, but not the only one, in determining the growth of the small plant and animal life of the sea, called plankton.

Most of the the open sea is a biological desert, an essentially barren area comprising 90 per cent of the ocean or nearly three-fourths of the earth's surface. Half the world's fish supply is produced in coastal waters and in a few offshore areas of comparably high fertility. The other half is produced in regions of upwelling water that total no more than 0.1 per cent of the ocean's surface. Thus, the most productive portions of the ocean—the coastal waters—are those which are the most polluted and where the marine life is the most susceptible to disturbances in the environnment.

Algal Blooms

Every lake or pond goes through a life cycle that may take from a few days up to thousands of years for the larger bodies of water. Each lake or pond faces the prospect of becoming more shallow with time as silt and organic matter are swept in by streams and run-off and deposited in the water. Shallow ends, often near the outlets of streams, tend to become marshes. Algae and other plants proliferate under these conditions and often appear as a green scum. We can see various stages in the life history of lakes, from those with fresh, clear, sparkling water to the shallow, growth-choked, scum-surfaced swamps on the verge of becoming dead lakes.

In lakes and slowly moving waters that have been enriched, algal blooms or dense plant growth accumulate along the shores and in shallow portions where they may form decaying masses and foul-smelling scums. Blue-green algae of several types are the most troublesome in water warmer than 20°C. Sometimes a toxin, as yet unidentified, is poisonous to livestock and wild animals.

Algal blooms are sometimes caused by conditions that are not related to added nutrients. Artificial warming of the water, such as that from cooling towers and condensers, may stimulate the growth of blue-green algae. The elimination of other aquatic flora, either mechanically or by the presence of some toxic substance, promotes the growth of bloom-forming algae by removing competition for the available nutrient supply. The release of organic nutrients by the decomposition of aquatic plants killed by herbicides may cause an increase in algal growth, and the absence of plankton-feeding organisms may favour excessive production of bloom.

Algae, that are the principal growth products of eutrophication, need at least fifteen elements to sustain growth. If any one of them is in short supply, it will be a limiting factor in the growth and development of the algae. However, many of the essential nutrients, particularly the so-called micro-nutrients, are almost always present in abundance in lake waters. Also, algae normally have access to an abundant supply of carbon, especially from carbon dioxide in the air and from carbonates dissolved in the water. (Some scientific investigators believe that a deficiency of carbon dioxide can become a limiting factor). Nutrients, which are utilized by the algal plants in large amounts, are apt to be the most important in determining the rate and degree of growth. The primary plant nutrients are nitrogen, phosphorus, and potassium. Of these three, nitrogen and phosphorus are in short supply naturally, and since some of the blue-green algae have nitrogen-fixing capabilities, phosphorus is in many cases the most important added nutrient in bringing about excessive growth.

There are several man-made sources of the nutrients that cause eutrophication. They include domestic, industrial, and agricultural wastes. Sewage is an important source of nitrogen and phosphorus as well as organic matter. Detergents are important contributors of phosphates. Agricultural fertilizers and livestock and poultry wastes are also major sources. Wastes from slaughter houses and food processing plants often end up in the waterways. Boats discharge a variety of pollutants including sanitary wastes.

Several points of attack are needed to solve the problems of water pollution. Better sewage treatment to remove nitrogen and phosphorus and more effective disposal of the organic material as sludge are needed on a large scale. Control of industrial and agricultural wastes would mitigate the problems. Since phosphate-containing detergents are a major contributor of phosphorus, the elimination of phosphates from detergents would go a long way toward retarding eutrophication. Because food and other agricultural products come from the land, the return of agricultural wastes to the land would complete their cycle.

Detergents

Huge quantities of detergents are used annually throughout the world. Most of it is flushed into the sewer systems from bathrooms, kitchens, laundries, factories, and other industrial establishments. Detergents consist of *surfactants** mixed with conditioning and water-softening agents, mainly polyphosphates.

The most commonly used surfactants now are the linear alkyl sulphonates, e.g., sodium dodecylbenzenesulphonate. These compounds are rapidly degraded by bacteria** and do not normally constitute a serious environmental hazard, unlike the branched chain alkyl compounds, e.g., tetrapropylene benzene sulphonate, which they replaced and which were resistant to bacterial degradation. However, all detergents have appreciable toxicity and, if released in large amounts, may reach concentrations which kill many organisms. This is most likely to occur when they are being used as dispersants to break up oil slicks on beaches or in natural waters. Even if the detergents do not kill the organisms directly, they may damage cell membranes sufficiently to allow other toxicants to act. Thus, the benefits of rapid oil dispersal by detergents must be weighed very carefully against the risks involved before their use is approved.

Detergents used for cleaning purposes contain compounds called builders. The most common builders used are polyphosphates. Sodium tripolyphosphate, the most widely used

**Surfactants* are surface active agents, substances having properties similar to soap.

**Materials that are decomposed by microorganisms are referred to as *biodegradable*.

phosphate builder in detergents, is one of a family of phosphates used extensively as water softeners.

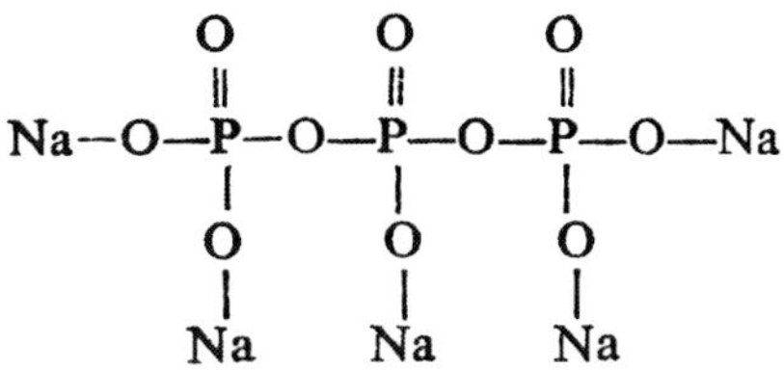

sodium tripolyphosphate

The value of the polyphosphates is in their ability to form soluble complexes with alkaline earth metal ions, mainly calcuim (Ca^{2+}) and magnesium (Mg^{2+}) found in hard waters. Compounds having this action are called *sequestering agents*. The polyphosphates sequester divalent ions such as calcium and magnesium so that they do not interfere with the cleaning action of the surfactant. Builders also make wash water alkaline, thus improving dirt removal.

The polyphosphates hydrolyse to phosphate when released into the natural environment, and this may contribute to excessive eutrophication, enrichment of the water with nutrients, accelerated growth of algae, oxygen depletion of the surface water, and impaired fish production. The ecological balance is destroyed.

However, substitutes for polyphosphates have all got drawbacks and it seems likely that their use will continue, but with care being taken to remove them from waste water before it is released. This can be done quite readily by precipitation with aluminium sulphate.

Industrial Pollution

The contribution of the chemical industry to human welfare is evident in the wide range of chemicals used in drugs, cosmetics, plastics, synthetic fibers, paints, cleansing agents, and many other kinds of consumer products. Unfortunately, most of the chemicals end up in the environment. A high proportion is sent down the sewers and ultimately into rivers, lakes, and oceans. Many of the chemicals are highly toxic to living organisms. Some of them find their way into the food chain, disrupting or impairing the natural biotic cycle and theatening populations of organisms with extinction. Natural waters from some rivers and lakes have become unfit for domestic use and some streams have become biologically barren.

Chemical Pollutants

The dependence of present-day society on chemical technology is almost beyond comprehension. Synthetic chemicals of every conceivable kind are part of our everyday life. Millions of chemical compounds are known and several thousand new ones are added each year. Thousands of synthetic organic compounds are used commercially, with 300 to 500 new ones being discovered annually.

Oceans do not have the rapid turnover that occurs in rivers, springs, and lakes. The oceans are more stable and maintain their chemical composition over periods of thousands of years. Injurious chemicals can build up in concentration over long periods of time unless they are decomposed. Poisonous metals are indestructible. Even some synthetic organic chemicals have remarkable stability, as, for example, DDT and many of the plastics.

Under the anaerobic conditions of silt and mud, chemical substances may be relatively free of oxidation and other chemical degradation as well as resistent to bacterial decomposition. But it cannot be assumed that chemical pollutants will remain indefinitely at the point of deposition. Some of the pollutants may become buried in the sediment near the coastline where they can be consumed by microorganisms and mud-inhabiting worms. Thus, chemicals and their degradation products are taken up in the food chain and passed from one species of organism to others that prey upon them. Some of the organisms may be carried by ocean currents; fish and other organisms may travel under their own power for great distances.

A large number of organic and inorganic substances have been identified in domestic sewage. Materials that are known to be potentially serious pollutants include sundry chemicals from domestic disposal, such as from household cleaners, medicines, and chemicals in food wastes; a wide variety of industrial chemical wastes, including petroleum products, phenols, solvents, chemical intermediates, metallic wastes, and by-products which may be highly toxic to living organisms; pesticides from agricultural use or industrial waste disposal; and radioactive chemicals, some of which have toxic properties lasting for thousands of years and whose fate in the food chain is largely speculative, though they are among the most toxic substances known. Many of these substances are discussed in subsequent chapters.

Industrial Disposal

Industry is the largest single user of water. Treatment of industrial waste waters is complicated by the presence of a wide variety of both inorganic and synthetic organic pollutants, many of which are not readily susceptible to biodegradation. Solvents, oils, plastics, plasticizers, metallic wastes, suspended solids, phenols, and various chemical derivatives of manufacturing processes are apt to be difficult to identify and impossible to remove without more advanced technology than we now possess. Some of the substances are known to be highly toxic to living organisms but are of unknown effect as environmental pollutants.

One method used to dispose of industrial toxins is to inject the waste water deep underground. However, disposal of toxic substances deep underground requires careful study of the possibility of irretrievably contaminating the underground water supplies because subsurface water may move for considerable distances underground. Another danger of underground disposal is that the pressure might cause slippage of deep layers of rock, resulting in earthquakes.

Water Pollution and Wild Life

There are many cases on record of the destruction of wildlife by polluted waters. Mass killing of fish was among the earliest and most dramatic results of indiscriminate pollution of water. The cases are too numerous to list. The contamination comes from diverse sources: municipal wastes of all kinds, especially sewage; agricultural wastes; industrial pollutants etc. It should be noted, however, that all mass die-offs of wildlife are *not* caused by man-made pollution. Ocean going tankers are a continuing source of oil pollution. The pollution caused by oil-spills destroys many forms of marine life.

Atmosphere in the Sea

Without dissolved carbon dioxide and oxygen, it would not be possible for the profusion of aquatic (fresh-water) and marine life to exist. The high solubility of carbon dioxide in water is of great importance in promoting photosynthesis by phytoplankton and, consequently, is beneficial to all organisms in the food chain. The relatively low concentration of oxygen in water compared to air is a disadvantage to aquatic and marine animals that depend on gills or similar arrangements for obtaining oxygen from water; it also accounts for the greater efficiency of air-breathing animals. However, some of the disadvantages are overcome by the intricate structure of gills and mechanisms for gas exchange. The increase in solubility of oxygen and carbon dioxide in water at low temperatures is favourable for the development of marine life. This explains why many of the world's most productive fisheries are in cold waters.

The bodies of water are in a state of equilibrium with the atmosphere and the earth because the solubility of the gases increases as the water temperature and salinity decrease. However, we do not find the same proportions of the gases in water as in the air, due primarily to differences in solubility. Pure water exposed to the atmosphere contains, *by volume*, about 2 per cent nitrogen, 1 per cent oxygen, and 0.05 per cent carbon dioxide at 0°C and normal atmospheric pressure. Compared to nitrogen and oxygen, the amount of carbon dioxide is proportionately much higher than in air. However, the concentration of oxygen and carbon dioxide in natural waters is highly variable. The differences are largely the result of the growth of algae and other aquatic organisms and the stagnation or mobility of the water.

Carbon Dioxide

Carbon dioxide in natural water comes primarily from two sources: from dissolved gas from the atmosphere, and from carbon dioxide liberated by organisms during respiration. The concentration varies from a fraction of a part per million to about 20 ppm.

Water is capable of absorbing carbon dioxide by forming carbonic acid:

$$\underset{\text{carbon dioxide}}{CO_2} + \underset{\text{water}}{H_2O} \rightleftharpoons \underset{\text{carbonic acid}}{H_2CO_3}$$

Because the reaction is reversible, the carbonic acid serves as a reservoir which makes some carbon dioxide available for plants when the dissolved gas nears depletion. Some of the carbonic acid dissociates into hydrogen ions and bicarbonate ions:

$$\underset{\text{carbonic acid}}{H_2CO_3} \rightleftharpoons \underset{\text{hydrogen}}{H^+} + \underset{\text{bicarbonate ion}}{HCO_3^-}$$

Some of the bicarbonate ions dissociate forming more hydrogen ions and carbonate ions:

$$\underset{\text{bicarbonate ion}}{HCO_3^-} \rightleftharpoons \underset{\text{hydrogen ion}}{H^+} + \underset{\text{carbonate ion}}{CO_3^{2-}}$$

The three reactions are in equilibrium. The equilibrium shifts to the right or left depending on the withdrawal or addition of ions by organisms, the acidity or alkalinity of the water, and the temperature. In the ocean the average concentration of bicarbonate ion is about 0.014 per cent.

Oxygen

In a body of water containing diversified biota, the oxygen content is in a condition of unstable equilibrium. Some of the oxygen is added by direct absorption from the air and some of it comes from the oxygen given off by plants during photosynthesis. Absorption of the oxygen from the atmosphere may be aided by surface agitation caused by wind and waves, but such action may also reverse the process of releasing oxygen to the atmosphere when the water becomes supersaturated. Oxygen is removed from the water by respiration of organisms and by the oxidation of dead organic material during its decomposition.

While aquatic plants absorb carbon dioxide from the water for use during photosynthesis, oxygen is released in the dissolved state; if the water is supersaturated with oxygen, gaseous oxygen is liberated. Oxygen is produced during the first (energy capture) phase of the photosynthetic process called *photolysis*. During this process the action of light energy on the chlorophyll indirectly brings about the decomposition of water into its components, hydrogen and oxygen. Since photolysis can occur only in the light, the production of oxygen is limited by the amount of sunlight available and can occur only in the region called the photic zone. The total oxygen productivity of an ecosystem* is determined by the amount of solar energy reaching the organisms, by the photosynthetic efficiency of the plants, and by physical factors such as temperature that affect the rate of photosynthesis. A covering of ice reduces both the penetration of light and the absorption of oxygen.

The air contains 250 milligrams per litre of oxygen at 20°C at atmospheric pressure. As the temperature goes down, the

*The ecosystem is the composite family of living organisms and their environment.

oxygen capacity of the water increases. Cold water at 4°C can hold 1.5 times as much oxygen as warm water at 25°C. The saturation level at 0°C in fresh water is about 15 ppm. The concentration may increase to two or three times as much when the rate of photosynthesis is high during daylight, and may fall far below the saturation level when photosynthesis declines at night.

Bacterial decomposition of organic matter may deplete the dissolved oxygen in stagnant deep water during the summer when there is an over-abundance of phytoplankton and other aquatic life. Similarly, the decomposition of organic matter in sewage wastes depletes the oxygen content. Organic pollution from either natural or human sources is sometimes expressed as the amount of oxygen needed by microorganisms to decompose the organic material. An evaluation is made by determining the amount of oxygen used up over a period of 5 days in the laboratory. This is called the B.O.D. (*biochemical oxygen demand*). The B.O.D. is an indication of the amount of organic material present in the water or wastewater.

A problem arises if there is a deficiency of oxygen to the extent that anaerobic conditions prevail. Whereas the aerobic oxidation of organic material results in generally harmless end products, anaerobic decomposition produces substances that are toxic and offensive.

Oxygen deficiencies are injurious to fish. A reduction in concentration of dissolved oxygen (D.O.) to about 3 milligrams per litre or less causes delayed hatching of eggs, reduced size and vigour of the embryos, deformities in the young, interference with the digestion of food, accelerated clotting of blood, reduced growth rate, poor utilization of food, impairment of swimming speed, and a decreased tolerance to toxicants. The lethal effect of low concentrations of dissolved oxygen appears to be enhanced by excessive carbon dioxide and ammonia as well as by cyanides, zinc, lead, copper, and cresols (cresylic acids).

Water and Health

Clean water was just as difficult to guarantee 2,000 years ago as it is today. The Romans did not have filtration plants, chlorination facilities, or similar modern equipment. They had to rely on local streams and springs or transport the water over long distances. The Roman aqueducts, a number of which are still in use today, provided clean water for cities. Where these sources were not available, wells had to be dug; and to assure good water, they were planned by reputed architects and builders.

Different water quality standards are required depending on the use for which the water is intended. The major beneficial uses are listed below:

— Domestic water supply, including drinking water.

— Industrial water supply, including cooling water.

— Agricultural water supply for irrigation.

— Livestock and wildlife water.

— Swimming, bathing, water skiting, surfing and other water sports.

— Boating and aesthetic enjoyment.

— Hydroelectric power and navigation.

— Waste disposal, dispersion, and assimilation.

Bacterial and Parasite Pollution

Bacterial contamination is the most common water-borne disease hazard the world over. In many parts of the world both bacterial and other types of organisms are ever-present threats. Typhoid fever and cholera periodically cause widespread illness and death. Various forms of gastroenteritis* are common. The protozoan *Entamoeba hystolitica* causes a type of dysentery that

*Gastroenteritis is inflamation of stomach and intestines.

is often serious. Parasitic worms of various kinds are transmitted by water. Eggs of tapeworms and roundworms are found in sewage. Pollution of streams by domestic and wild animals may be a source of contamination. Several species of flukes, all of which are parasitic worms, cause serious diseases in different parts of the world.

An early method of determining the microbial purity of water was to take a reading of its "keeping power." Water that was free of bacteria could be stored for a long time, while contaminated water would tend to develop growths and undergo visible changes. Increasingly widespread use of the microscope made it possible to establish standards based on the numbers and kinds of organisms that could be found in the water. Later, bacteriological tests were devised that were reliable and specific for the detection and identification of microorganisms.

Examination of water for each of the various kinds of pathogenic microorganisms is slow, unwieldly, and costly. Therefore, bacteriological tests are used to detect the presence of bacteria that are presumed to be of faecal origin and thus indicative of sewage contamination. When indications of faecal pollution are found, the water is considered to be potentially dangerous to health. The bacteria used as indicators are described as *coliform* organisms, a group which includes diverse microorganisms, the best known of which are strains of *Escherichia coli* and *Enterobacter aerogenes*, which are usually, though not always, of faecal origin. Coliforms may also be nonfecal types that thrive on organic material of vegetable origin.

Coliform bacteria may be removed by chlorination and filtration. Even when this is done, it is important to identify coliform bacteria in the raw water because their presence is an indication of sewage pollution. Sewage may carry other bacteria that are resistant to treatment in addition to viruses, organic matter, and chemical contaminants having uncertain, but dangerous effects on human health.

Chemical Quality

Water untouched by man's work is not necessarily clean by nature. Each day the rivers carry millions of tons of mud and silt and sweeps the sediment into the seas. Even water near natural points of origin, such as that issuing from springs, is not necessarily chemically pure. Many waters contain harmful quantites of flouride. Arsenic, and heavy metal ions, high concentration of salts of various kinds and sulphuric acid are naturally present in some waters (as in the Lemonade Spings of New Mexico).

Human Waste

Part of the living process is to get rid of unwanted material. Otherwise organisms die of the toxins in their own waste. The biological process for ridding the body of wastes is called *excretion* and takes place in some manner in all living creatures. Man and other vertebrates have several mechanisms for excretion. In man, the skin eliminates unwanted water, salts, and carbon dioxide; the lungs expel carbon dioxide and water; the liver drains bile constituents into the alimentary canal; the kidneys purify the blood and expel the wastes through the urinary system; the intestinal tract gets rid of salts, minerals, fats, and indigestable material rejected as unsuitable for food or metabolic processes.

The excreta of the alimentary canal are called *faeces*. They consist primarily of intestinal bacteria, which comprise the most bulky portion of human waste. The urine, of which about 1.5 litres are excreted daily by a human adult, is mostly water with dissolved nitrogenous wastes and salts. The urine also carries away foreign substances such as drugs and other toxins.

Disposal of the Wastes of Life

When no more than 100,000 people populated the earth during the Ice Age, human excretion was not a problem.

It was accomplished by walking to the woods or the rocks and leaving the waste for disposal by coprophagous (dung-eating) organisms. Nature has several answers to the population problem. One effect of overcongestion is that the waste products eventually become so concentrated that many of the organisms die out from toxins, disease, or abnormal behavior. Sometimes they become cannibalistic or otherwise self-destructive. Some of them may survive by escaping to less congested areas. Civilization, with its concentration of humans living in cities, aggravates the problem of waste disposal beyond that which can be solved by natural means.

The people of the Harappan culture of the Indus Valley, circa 2500 B. C., were perhaps the first to become experts in sanitation. Household water supplies, bathing facilities, and drainage systems were widespread. There were bathrooms with waterproof floors. Latrines with seats were built into some houses. There were sloping or stepped channels through the wall to either a pottery receptable or brick drain outside. At Mohenjo-Daro each house had a drain which ran into a central sewer system under the street which fed into cesspools. Manhole covers of brick were installed to allow for repair and cleaning.

The ancient cities of Sumer and Babylon seem to have had similar hygienic devices. In the city of Akhenaton in Egypt, built during the Eighteenth Dynasty (1580—1340 B.C.), upper-class houses were equipped with bathrooms, drains and stands for holding jars of drinking and washing water.

The use of waste for productive purposes is an ancient solution to the disposal problem. Animal dung has been used as fertilizer for thousands of years. In the Orient* human dung has been used as fertilizer for centuries and is today a convenient answer to the dual problems of disposal and food production. The practice is not aesthetically or economically satisfactory in an industrial society, although dried sludge is produced by some modern sewage treatment plants as a by-product and sold as horticultural fertilizer.

*The *Orient* refers to the part of world that includes the Indian sub-continent, China and Japan.

It were perhaps aesthetic reasons, rather than hygienic reasons which stimulated the construction of sanitary facilities. Cesspools were dug everywhere and many of them were prolific breeders of contamination because of seepage into nearby wells. As a result, typhoid fever and cholera epidemics were rampant. With the passage of time the methods were improved. London installed a sewer system in 1865 and many English cities followed the example. Similar advances were made by American cities.

The composition of sewage is complex, and it differs depending upon the sources, the type of treatment or lack of it, and whether there is an admixture of storm drainage with industrial waste. Unfortunately, analyses and inventory of waste constituents, a necessary step for rational planning of water pollution control, have been thoroughly done in only a few cases. Inorganic constituent, such as the salts of sodium, potassium, calcium, magnesium, and ammonium, together with chloride, nitrate, bicarbonate, sulphate, and phosphate ions are commonly found in abundance. Organic chemical compounds in waste have been poorly investigated, with the exception of some extensive work on pesticides and detergents and, to a more limited extent, on organic acids and phenols. A large number of organic and inorganic substances have been identified in domestic sewage. Many of them are probably breakdown products of excreta.

REFERENCES

Anon. 1970, Drinking water—Is it Drinkable? *Env. Sci. & Tech.* 4 (10) 811—13.

Varshney, C. K. (Ed). 1985. *Water Pollution and Management.* Wiley Eastern Limited, New Delhi.

13

Toxic Metals

In geological terms, trace elements are defined as those occurring at 1000 ppm or less in the earth's crust. Only twelve elements do not come within this classification, *i.e.*, oxygen, silicon, aluminium, iron, calcium, sodium, potassium, magnesium, titanium, hydrogen, phosphorus and manganese. The trace metals may be divided into those that are 'heavy', with densities greater than 5 g cm^{-3}, and those that are 'light', with densities less than 5 g cm^{-3}. Many trace metals are essential at low concentrations for normal life. Since most organisms are not adapted to deal with high concentrations of trace metal, the problem of metal toxicity arises.

Excessive levels of trace metals may occur naturally as a result of normal geological phenomena such as ore formation. Weathering of rocks, leaching or, in the case of mercury, degassing may make these metals available to the biosphere. Man releases more of the metals by burning fossil fuels, mining, smelting, discharging industrial, agricultural and domestic waste, and by deliberate environmental application in pesticides. Once made available to the environment, metals are not usually removed rapidly, nor are they readily detoxified by metabolic activity. As a result they accumulate. Thus, their release into the environment must be carefully monitored and controlled.

The EPA has defined beryllium (a light trace metal) and mercury (a heavy trace metal) as hazardous, meaning that slight exposure can endanger human health. Nine other metals have been defined as potentially hazardous and they must be kept under review. They are barium, cadmium, copper, lead, manganese, nickel, tin, vanadium and zinc. Of these, all but manganese are trace metals and all but barium are heavy metals. Toxic effects of some metals are summarised in Table 13.1.

TABLE 13.1 **Toxic Effects of Trace Metals**

Metal	*Sources*	*Toxic Effects*
Antimony	Industry	Shortening of life spans in rats.
Beryllium	Coal, industry	Most toxic, accumulates in the lungs to produce beryl-liosis, a serious disease.
Bismuth	Coal	Kidney and liver damage in large doses.
Cadmium	Coal, zinc mining, tobacco smoke	Cardiovascular disease and hypertension in humans suspected; interferes with Zn and Cu metabolism.
Lead	Auto exhaust	Brain damage, convulsions, behavioural disorders, death.
Mercury	Coal, industrial fungicides, electrical appliances.	Nerve damage, death.
Nickel	Diesel oil, coal, tobacco smoke, chemicals, catalysts steel and non-ferrous alloys	Carcinogenic properties in animals, and in human when inhaled as the carbonyl, Ni $(CO)_4$
Tin	Coal, iron and steel production, tin plating	Low order of toxicity, decreased life span in rats and mice.

Source: Schroeder, H.A. 1970. *Arch. Environ. Health*, **20**, 803.

Lead

Lead is found in our water, food, and air. Most of the atmospheric lead comes from antiknock compounds added to automobile fuel. Typical samples contain between 2 to 4g of lead per litre with an average of about 2.8g.

On an average 70 to 80 per cent of the lead in fuel is emitted by the automobiles as particulate. The emission is less at low-speed driving and is more on highways. Most of the lead particulate from automobiles is less than 2μ in diameter and thus does not settle out of the air rapidly. Inorganic lead acts as an agent to cause a variety of effects on human health including liver and kidney damage, gastrointestinal damage, mental health effects in children, and abnormalities in fertility and pregnancy. Adults with less than 80 μg/100g of blood, and urine lead less than 150 μg/litre are considered to be within normal limits. Fortunately lead is removed from the body primarily with the urine. People who have an additional occupational exposure can build up high concentrations of lead. The following conclusions are generally agreed upon:

1. Men have greater levels of lead in their blood than women,
2. Urban residents have greater levels than rural residents and
3. Smokers have greater levels than non-smokers.

Beryllium

Beryllium is released during the burning of coal, but the main environmental hazard is to those working in industries where beryllium is produced or used, e.g., the manufacture of nuclear reactors, aircraft and rockets. In humans, beryllium has been shown to damage skin and mucous membrane. It accumulates in the lungs where it causes beryllium disease (berylliosis).

It may cause cancer in lungs and bone marrow. It is not excreted from mammalian tissue and, therefore, its effects are cumulative. At the biochemical level, beryllium competes with magnesium for enzyme sites and has been shown to inhibit DNA polymerase, thymidine kinase and alkaline phosphatase.

Mercury

Mercury is concentrated in various ores, the principal one being cinnabar (HgS). It has been mined since 700 BC and is currently used industrially in three forms: as the metal, in organic compounds and in inorganic compounds. The greatest use of mercury is in the production of electrical apparatus. The second greatest use is in the chloro-alkali industry, which produces chlorine and caustic soda by electrolysis of sodium chloride solution using mercury as the cathode of the electrolysis cell. The third greatest use worldwide is in fungicides.

Nearly all the mercury used by man eventually enters the natural environment. To this may be added amounts released during the production of mercury and other metals, from coal burning and from weathering of rocks.

All forms of mercury are potentially toxic but the toxicities vary considerably. The least toxic are the inorganic mercury compounds. They are not readily absorbed from the gastro-intestinal tract. Once absorbed they may accumulate in the liver and kidney but normally they are excreted quite rapidly in the urine. It is worth noting that mercury amalgam has long been used in dentistry to fill teeth without any toxic effects being noted. Mercury vapour is the most hazardous of the inorganic forms because it can diffuse through the lungs into the blood and then into the brain, where serious damage can occur. Arylmercurials are as toxic as the inorganic forms since they are readily broken down to inorganic derivatives in the tissues. Alkylmercurials are the most toxic mercury compounds so far studied. They are fairly stable and have long retention times in the tissues. Therefore, they readily accumulate to high concentrations. Their lipid solubility gives them an affinity for nervous tissue which accounts for many of their harmful effects. In addition, they have been reported to cause abnormalities in cell division and to increase the frequency of chromosome breakages. Some of these abnormalities may be due to combination of mercurials with sulphydryl groups. Inhibition of enzymes in this way has been demonstrated frequently. So far, no effective treatment for mercury poisoning has been developed, though British antilewisite (BAL) or calcium ethylenediaminetetraacetate (Ca_2EDTA) may have some alleviating effect.

Unfortunately, in assessing the risk from mercury in a particular environment, it is not enough to know the form in which it entered that environment because various transformations can take place. Probably the most serious of these is the transformation of metallic mercury to methyl and dimethyl derivatives by anaerobic micro-organisms, especially *Clostridium cochlearium*, in aquatic sediments. This may also occur in decaying fish. Under aerobic conditions, this transformation can be brought about by *Pseudomonas* spp, and by the fungus, *Neurospora crassa*. In essence, it represents conversion of the least toxic form of mercury to the most toxic. Other transformations which can be brought about by bacteria are the following: phenyl, ethyl and methyl mercury can be reduced to elemental mercury and benzene, ethane and methane, respectively; phenylmercuricacetate can be converted aerobically to elemental mercury and diphenyl mercury; mercuric ions can be reduced to elemental mercury. Not all the transformations observed are biological. Under alkaline conditions, methyl mercury converts to the more volatile dimethyl mercury. Under oxidizing conditions, in the presence of ultraviolet light, phenyl mercury, alkoxyalkyl mercury and alkyl mercury may break down to give inorganic mercury. Under anaerobic conditions, mercuric ions may combine with hydrogen sulphide to form poorly-soluble mercuric sulphide. With subsequent aeration, mercuric sulphide can be converted to the soluble sulphate which may then be methylated biologically.

Most macro-organisms are relatively insensitive to mercury and its derivatives. Nitrogen fixing bacteria in the soil require levels of abouc 100 ppm before they are adversely affected. Normal soil levels are between 0.005 and 1 ppm. However, marine and freshwater phytoplankton, especially diatoms, are very sensitive to organomercurial fungicides and as little as 0.001 ppm may reduce their photosynthetic efficiency. Many algae and other plants have the ability to absorb and concentrate mercury from the surrounding environment. Droplets of elemental mercury have been found in chickweed. Such high concentrations may cause mitotic disturbances and kill the plants. Fortunately, most agricultural plants do not seem to absorb much mercury. Animals tend to accumulate mercury through their food. Pike* can accumulate a concentration of mercury 3000 times higher than that in the water in which they live. Tuna and Swordfish show the same ability. Similar observations have been made on predatory birds. Seed-eating birds accumulate mercury where alkylmercury seed dressing are being used. Much less is accumulated where alkoxylakyl compounds are used, and negligible quantities where inorganic mercury compounds are used.

The effects of mercury and its derivatives on man deserve special consideration because it was mainly these that caused concern about the effects of heavy metals released into the environment in large amounts. The first serious incident to come to light occurred at Minamata Bay in Japan. In this case, comparatively nontoxic inorganic mercury along with some methyl mercury was released in effluent by a chemical factory using mercuric sulphate catalysts in acetaldehyde production. The effluent entered a river running into Minamata Bay. In the sediments, the inorganic mercury was converted to methyl mercury. This accumulated in shellfish and fish which were eaten by the local inhabitants. Consequently by 1975, 115 people had died and many were left paralysed for life. Others

*A *pike* is a large fish that lives in rivers and lakes, and that eats other fish.

suffered impairment of vision and hearing and other neurological symptoms. Prenatal poisoning of the foetus was observed even in the absence of symptoms in the mother. Since the Minamata Bay incident, another has occurred around the Agana River, Niigata, Japan. This led to 23 deaths. In both these cases, many domestic animals, fish, shellfish and seabirds were affected.

Mercury poisoning of human beings has also been caused by the consumption of food containing high concentrations derived from alkylmercury agricultural seed treatments used to prevent seed-borne disease. In Iraq, treated seeds, intended for planting, were used to make bread. Thousands were poisoned and hundreds died. In the USA, cattle fed on treated grain were slaughtered for human consumption. Again, severe poisoning resulted. Consequently a number of countries have now banned the use of alkylmercurial seed treatments. A committee of experts constituted by the Food and Agriculture Organization (FAO) and World Health Organization (WHO) has recommended that the use of alkylmercurials should be restricted to stocks of cereal seed used for plant breeding or seed production, and never permitted for treatment of cereal seed for export for the production of food.

Despite the major incidents referred to above, it seems fairly certain that the average person is at no great risk from exposure to mercury. The normal dietary intake is well below what is thought to be the tolerable limit of 0.3 mg per person per week, of which not more than 0.2 mg should be in the methylated form, according to WHO. For most people, the chances of appreciable exposure from air, pesticides or pharmaceuticals are very limited. Where there is a risk of occupational exposure, this should be minimized by appropriate precautions and medical screening. With regard to the general environment, there is still a need to know more about the concentration and distribution of mercury, especially with regard to those areas where localized high concentrations do exist.

Arsenic

Arsenic occurs widely in nature. It is used in alloys, pesticides, wood preservatives and some medical preparations. It was formerly used in paint pigments, but this use discontinued when it was found that, under damp conditions, moulds converted the arsenic to the highly toxic gases, arsine, AsH_3, and trimethyl arsine, $As(CH_3)_3$. Arsenic is a cumulative poison, causing vomiting and abdominal pains prior to death. It may also cause dermatitis and bronchitis, and may be carcinogenic to tissues of the mouth, oesophagus, larynx and bladder. At the cell level, it can uncouple oxidative phosphorylation and compete with phosphorus in metabolic reactions.

Arsenic is concentrated by organisms exposed to it and accumulates along food chains. Accumulation in fish seems to be favoured by increasing salinity. Anand (1978) has reported arsenic pollution loads in fish sold in Bombay. Crabs and lobsters have been especially noted for accumulating high concentrations, but no cases of human poisoning seem to have arisen from this.

Barium

The most common naturally occurring form of barium is barium sulphate, sometimes called barite or baryta. Barium derivatives are used as fillers for rubber, linoleum, etc., as pigments in paint, in glass manufacture, in the ceramic industry and in various other industrial applications. These applications constitute the main source of risk, primarily to workers in the appropriate industries. Barium salts are highly toxic. They cause vomiting and diarrhoea, which may be associated with stomach, intestinal and kidney haemorrhage. They also affect the central nervous system, causing convulsions and have been implicated as a cause of pneumoconiosis. Despite the known toxicity of barium salts, barium sulphate is used to coat the alimentary tract for X-ray photographs since barium absorbs X-rays strongly and thus increases the contrast. The insolubility of barium sulphate minimizes its toxicity, and makes its use acceptable.

Cadmium

Cadmium is a normal constituent of soil and water at low concentrations. It is usually mined and extracted from zinc ores, especially zinc sulphide. Industrially, cadmium is used as an anti-friction agent, as a rust proofer and in alloys. It is also used in semiconductors, control rods for nuclear reactors, electroplating bases, PVC manufacture and batteries. In the environment, cadmium is dangerous bacause many plants and animals absorb it efficiently and concentrate it within their tissues. Normally, however, retention from food by mammals is low but absorption is increased if the mammals are on a low calcium diet. Once absorbed, cadmium associates with the low molecular weight protein, metallothionein, and accumulates in the kidneys, liver and reproductive organs. Very small dose can cause vomiting, diarrhoea and colitis. Continuous exposure to cadmium causes hypertension, heart enlargement and premature death. There is some evidence suggesting that cadmium can induce chromosome abnormalities and may exert a carcinogenic effect on the lungs.

In 1955 there was a cadmium poisoning incident in Northern Japan caused by the accumulation of cadmium in rice and soya beans. This was characterized by extreme lumbago and skeletal collapse, apparently due to increasing bone porosity caused by inhibition of bone repair mechanisms. This was called Itai-itai (ouch-ouch) disease because of the severe pains associated with it. Apart from this, the principal hazard again is to those working in the industries which use cadmium where appropriate precautions must be taken.

The occurrence of cadmium in river waters in different parts of the world is reported to be in the range 1 to 100 mg Cd per kg of the sample. Srivastava and Jaiswal (1989) have reported that whereas cadmium retards growth in the water plant *Spirodela polyrrhiza* it induces the formation of vegetatively propagating organ called *turion*. This is a distinct adaptive strategy to ensure successful recurrence of the species through turions after the stress is over in the water environment. They have noted that cadmium toxicity in this species is in the concentration range 0.005 ppm to 2.00 ppm and the adverse effects are on pigmentation and protein concentration.

The concentration of Cd in sea water on an average is 0.15 mg per litre (Goldberg, 1976). Anand (1978) has reported 16-17 mg Cd per kg of sample in fish sold in Bombay which is of the same order reported in other parts of the world. The toxic levels are 200mg Cd per litre of water.

Chromium

Like most of the metals discussed here, chromium is widely distributed. For most organisms, it is essential as a micronutrient in trace quantities for fat and carbohydrate metabolism. In industry, it is used in making steel alloys, in chromium plating and in leather tanning. Chromates are water soluble and can poison sewage treatment processes. The chromium ion can exist in four valency states; Cr^{2+}, Cr^{3+}, Cr^{5+} and Cr^{6+}. Of these, the hexavalent ion is the most toxic and it should be reduced to the trivalent state to form insoluble products before chromium waste is released into the environment. Hexavalent chromium has been implicated in poisoning in Japan. In this case, aerosols from chromium refining plants appear to have affected a considerable number of people causing lung cancer. Besides this, it has been shown that chromates act as irritants to the eyes, nose and throat, and chronic exposure may lead to liver and kidney damage. A characteristic effect on human beings is the appearance of perforations in the nasal septum. At the cell level it appears that hexavalent chromium may cause chromosome abnormalities. Chromium is particularly dangerous because it accumulates in many organisms. Some aquatic algae have been shown to concentrate it 4000 times above the level of their immediate environment.

The maximum permissible limit of Cr in drinking water as recommended by the W.H.O, is 0.05 mg per litre. Rai and Raizada (1987 and 1988) have studied biochemically the effect of nickel, silver, chromium and lead on toxicity and its regulatory mechanisms on the nitrogen fixing blue geen algae particularly *Nostoc muscorum* when treated singly and in different combinations.

Copper

Copper is one of the most abundant trace metals. It is widely used in its metallic state, either in the pure form or in alloys. For almost all organisms it is an essential micronutrient. It may occur in very high concentrations in water, sediments and biota in some localized areas as a result of mining activities, of intensive use of copper pellets in pig rearing, or of the application of copper fungicides. However, there is no evidence of food chain magnification. Hence, most toxic effects are due to immediate exposure to the element. All organisms are harmed by excessive concentrations, which may be as low as 0.5 ppm for algae. Most fish are killed by a few parts per million (Lopez and Lee, 1977). In higher animals brain damage is a characteristic feature of copper poisoning.

Pillai (1985) has reported the presence of copper in Periyar river water, sediments and fish from South India. The upstream region water contained 0.05 to 0.1 ppm copper and the industrial area 0.025 to 0.3 ppm while the sediment of the respective zone had 1 to 100 and 5 to 100 ppm copper. Fish contained about 0.7 ppm. Copper occurs in water in different forms. Humic acid, present in natural waters is known to form complex with copper. Copper complexes are less toxic to water plants and animals in the pH range 6—8. Copper-humic acid form is common in freshwaters while $Cu(OH)_2$ and $CuCO_3$ are characteristic forms in sea water.

Iron

Iron is the fourth most abundant element in the earth's crust. Its greatest use is for structural iron and steel, but it is also used for making dyes and abrasives. It is an essential micronutrient required in trace quantities for the normal metabolism of plants and animals. It is a constituent of cytochromes and nonheme iron proteins involved in photosythesis, N_2 fixation and respiratory linked dehydrogenases (Noggle and Fritz, 1986) Ingestion of excessive amounts may result in the inhibition of activity of many enzymes. The amounts consumed must be very large because only a small proportion of all iron ingested is absorbed from the gastro-intestinal tract. Inhalation of iron dust can cause benign pneumoconiosis and can enhance harmful effects of sulphur-dioxide and various carcinogens.

Many streams are poisoned by high levels of iron in acid mine drainage. Pyrite, iron sulphide, is often found in close association with coal deposits. Upon exposure to moisture and atmospheric oxygen, the ferrous iron is oxidized to the ferric state, a reaction which is frequently accelerated by bacteria of the Thiobacillus—Ferrobacillus group. The ferric iron can then react with sulphide in the presence of water to produce sulphuric acid, or react directly with water to produce a yellow, flocculent mass of ferric hydroxide. Besides being acidic, water affected in this way becomes deficient in oxygen. Such poisoning of streams is reckoned to be one of the main causes of fish kill. Although particularly associated with mining, streams running through iron-laden strata may become poisoned spontaneously.

Manganese

Manganese ocurs widely in nature and is of considerable importance in the manufacture of steel. Biologically, it is an essential micronutrient for most organisms. It is required for activity of some dehydrogenases, decarboxylases, kinases, oxidases, peroxidases etc. It is required for photosynthetic evolution of oxygen. This metal is not a serious pollutant as in most waters its concentration is quite low ranging from 0.005 to 1 mg per litre. Potassium permanganate is used in very small doses to disinfect well water particularly in Indian villages. Excessive amounts of manganese affects animals adversely, causing cramps, tremors and hallucinations, manganic pneumonia and renal degeneration.

Nickel

Nickel is used in various forms for nickel plating, as a catalyst, as a mordant and in ceramic glazes etc. Again it is a micronutrient for most organisms but excessive quantities have toxic effects. In animals these include dermatitis and respiratory disorders, including lung cancer following inhalation. Amongst enzymes inhibited are cytochrome oxidase, isocitrate dehydrogenase and maleic dehydrogenase. A particularly poisonous derivative of nickel is nickel tetracarbonyl, $Ni(CO)_4$.

Selenium

Strictly speaking selenium is not a metal, though it has certain metallic properties. It is a member of the sulphur-group, produced as a byproduct of the extraction of copper, nickel, gold and silver ores. It is used in electronics, and in paints and rubber compounds, for most organisms it is an essential micronutrient but it can be toxic at very low concentrations. The maximum permissible concentration in drinking water is 0.01 ppm. Poisoning of livestock has occurred where cattle have eaten plants of the Brassica family which have taken in selenium and incorporated it in cysteine and methionine in place of sulphur. Selenium itself irritates the eyes, nose, throat and respiratory tract. It can cause cancer of the liver, pneumonia, liver and kidney degeneration and gastro-intestinal disturbances.

Tin

Tin is widely used mainly for making tinplate and various alloys and compounds. It is an essential micronutrient and the main cause for concern regarding its toxicity has been the development of trialkyl-tin and triaryl-tin compounds having powerful biocidal properties. These are used on growing crops as fungicides and insecticides. They are also used as antimicrobial agents and in marine anti-fouling paints. They can damage crops and, in animals, accumulate in the central nervous system with harmful effects.

Vanadium

Vanadium is widely distributed. It is used as an alloying element for steel and iron, in making oxidation catalysts and in colouring agents used in the ceramic industry. Large amounts enter the atmosphere from the burning of some petroleums. It is an essential micronutrient and may be accumulated by some marine organisms to concentrations many times higher than those in the surrounding water. Excessive levels in animals inhibit tissue oxidation and synthesis of cholesterol, phospbo-lipids and other lipids, and amino acids. Such levels may also cause precipitation of serum proteins.

Zinc

Zinc makes up only 0.004% of the earth's crust. Its most important use is as a protective coating on other metals, particularly in galvanizing iron and steel. It is an essential micro-nutrient and an essential constituent of alcohol dehydrogenase, glutamic dehydrogenase, lactic dehydrogenase, carbonic anhydrase, alkaline phosphatase, carboxy peptidase B, and other enzymes (Noggle and Fritz, 1986). In most natural waters zinc is found in traces (less than 1 mg per litre *i.e.*, well within the safe limits). Concentrations above 5 mg per litre causes disagreeable taste. In drinking water the level of zinc usually ranges from 0.005 to 1 ppm or mg per litre, but in certain regions it may exceed upto 7.0 mg per litre. Zinc is generally regarded as one of the less hazardous elements, though its toxicity may be enhanced by the presence of arsenic, lead, cadmium and antimony, as impurities. Toxic effects have been observed from the inhalation of fumes from galvanizing baths. The 'zinc fever' produced is characterized by chills, fever and nausea. Removal from the fumes leads to complete recovery. Zinc chloride fumes have sometimes caused fatal oedema of the lungs. Zinc or galvanized containers are not recommended for food storage but are acceptable for storing drinking water. This is because acidic foods can dissolve enough zinc from the container to cause poisoning. A factor which serves to minimize the risk of zinc poisoning is that it appears to be lost along food chains, unlike methyl mercury or cadmium, for example, which accumulate.

REFERENCES

Anand, S.J.S. 1978. Determination of Mercury, Arsenic and Cadmium in fish by neutron activation. *J. Radio. Anal. Chem.* **44**, 101-107.

Goldberg, E.D. 1976. *The Wealth of the Oceans*, The UNESCO Press Paris, 172.

Lopez, J.N. and G.F. Lee. 1977. Environmental Chemistry of Copper in Torch Lake, Michigan, *Water, Air* and *Soil Pollution*, **8**, 373-385.

Noggle, G.R. and G.J. Fritz 1986. *Introductory Plant Physiology*, Prentice-Hall of India Pvt. Ltd., New Delhi.

Pillai, K.C. 1985. Heavy metals in aquatic environment. In C.K. Varshney (Ed.) 1985. *Water Pollution and Management*, Wiley Eastern Limited, New Delhi.

Rai, L.C. and M. Raizada. 1987. Toxicity of nickel and silver ions to *Nostoc muscorum*: Interaction with ascorbic acid, glutathione and sulphur containing amino acids. *Ecotoxiology and Environmental Safety*, **14**, 12-20.

Rai, L.C. and M. Raizada, 1988. Impact of Chromium and lead on *Nostoc muscorum*: Regulation of toxicity of ascorbic acid, glutathione and sulphur containing amino acids. *Ecotoxiology and Environmental Safety*, **15**, 21-32.

Srivastava, Alka and V.S. Jaiswal. 1989. Effect of cadmium on turion formation and germination of *Spirodela polyrrhiza L.J. Plant Physiology*, **134**, 385-387.

14

Pesticides and Herbicides

Pesticides are substances used to control organisms which may adversely affect public health, or organisms which attack food and other materials essential to mankind. Such organisms include vermin, insects, nematodes and fungi. Herbicides are substances used to eliminate unwanted plants in agriculture and horticulture. Herbicides have also been used in jungle warfare in Vietnam to destroy tropical forests.

Pesticides and herbicides have made possible great increases in food production and improvements in human health. Where harm has occurred, this has largely been due to ignorance of the properties of these chemicals and misuse, *i.e.*, wrong quantity, application or timing. Since it is impossible to know every property of every pesticide or herbicide, it is inevitable that some environmental damage will occasionally result from their use. However, this is no argument against their use, so long as the benefits can be clearly shown to outweigh the damage. The important thing is to be aware that there is a finite risk of damage, to minimize it by appropriate toxicity testing of chemicals before they are applied on a large scale and to ensure that the application is carried out correctly. When pesticides are applied on a large scale, a constant watch must be kept for environmental damage so that corrective steps may be taken as soon as possible.

Everyday cases of poisoning by pesticides in developing countries are reported. In 1975, hundreds of people in Karnataka's Shimoga district in India were struck by a mysterious attack of arthritis which wastes away limbs and brings about dwarfism. Studies indicate that pesticides that were sprayed in fields were ingested by crabs which, in turn, were eaten by farmers, poisoning them as well. According to an WHO estimate about 750,000 people are poisoned by pesticides every year, 14000 of which are fatal. Three-fourths of these occur in the third world. The use of pesticides has been increasing at snowballing rates. In India, pesticide consumption rose from 2000 tonnes in 1955 to 80,000 tonnes in 1984—a 40-fold increase. By the end of the current year, the estimated consumption is projected at 100,000 million tonnes.

The scale of ecological havoc being wrought by pesticides is already alarming. When a pesticide is sprayed, much of it reaches the soil. Water becomes contaminated from direct application or run-off. Minute quantities become airborne. Many chemicals become metabolized by animals that eat plants sprayed with pesticides. We, thus, find these toxic chemicals into the human food chain.

Misuse of pesticides can often lead to major problems. Wearing gloves and face masks are a must as pesticides enter the skin through feet and hands if left unprotected. This could lead to beefy dermatitis. Containers used to store pesticides should never be used for storing grain.

Cosmetic pesticides are sprayed indiscriminately on fruits and vegetables in major cities in India to improve the 'looks'. Methyl parathion on cauliflower gives an extra white look; ladyfinger ('bhindi') is dipped in Copper sulphate to look greener. Washing vegetables with a 5 per cent solution of edible soda or vinegar, followed by a rinse of water will remove residues.

Insecticides

Chlorinated Hydrocarbons

Chlorinated hydrocarbons (Fig. 14.1) are the most prevalent pesticides in the environment because of their wide use and persistence. The most persistent of these compounds are DDT and derivatives such as DDD and DDE. DDE has an environmental half life of ten years or more, DDE can persist for decades. Almost as persistent are lindane (BHC) and heptachlor. Less persistent are aldrin and dieldrin, but even these require $2^1/_2$ years in the soil for 95 per cent degradation. All of the compounds mentioned are currently in use throughout the world, though DDT, aldrin and dieldrin have been banned for most purposes in the United States, and their use has been discouraged in the United Kingdom and a number of other countries. Reaction against the use of these compounds has followed awareness of their toxicity to nontarget organisms and the discovery that some target insect populations were becoming resistant. However, substantial quantities of chlorinated hydrocarbons are still used in wood preservation and where there is, at present, no satisfactory substitute. The most extensive use of DDT now occurs in tropical countries because it is cheap, persistent, generally effective and with minimum harm to human beings. Until a suitable alternative is available, its use will continue. At present, possible substitutes are much more expensive, less presistent and more toxic to human beings.

Most of the problems caused by effects of DDT on nontarget organisms are the result of its being used in excessive amounts. The excess finds its way into ponds, lakes, rivers and, ultimately, the sea. In some cases, DDT is applied directly to fresh water insect habitats; in others, accidental spraying to such habitats may occur. Otherwise, DDT may enter waterways in surface run-off, or be washed out of the atmosphere in rain or snow. Despite this, the concentration in natural waters is low. However, many plants and animals tend to accumulate DDT and its more persistent derivative DDE (Fig. 14.2). For example, oysters can concentrate DDT from 1 ppb in seawater

Structure	*Trivial name*	*Scientific name*
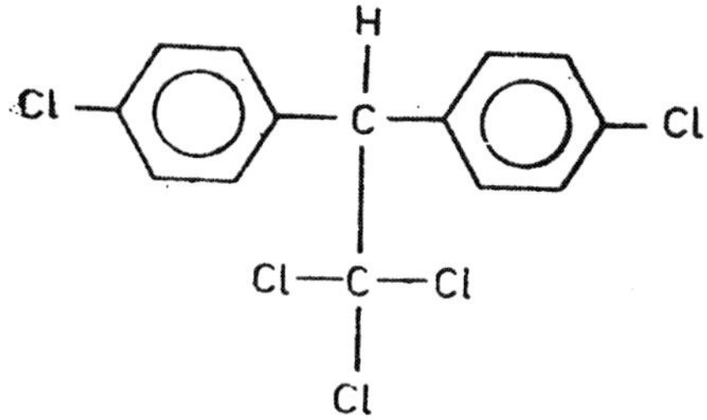	DDT	1,1,1-Trichloro-2,2-bis (*p*-chlorophenyl) ethane
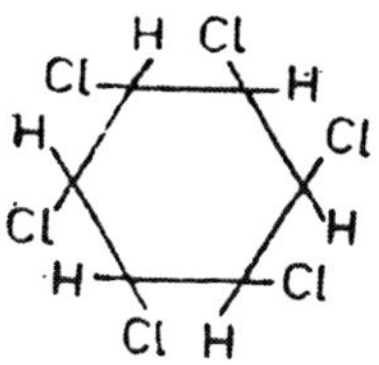	Lindane (BHC)	1,2,3,4,5,6-hexachloro-cyclohexane (γ-isomer)
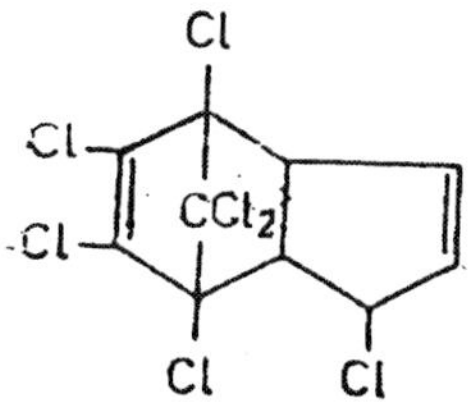	Heptachlor	1,4,5,6,7,8,8-heptachloro-3α, 4,7,7α-tetrahydro-4,7-endomethanoinde
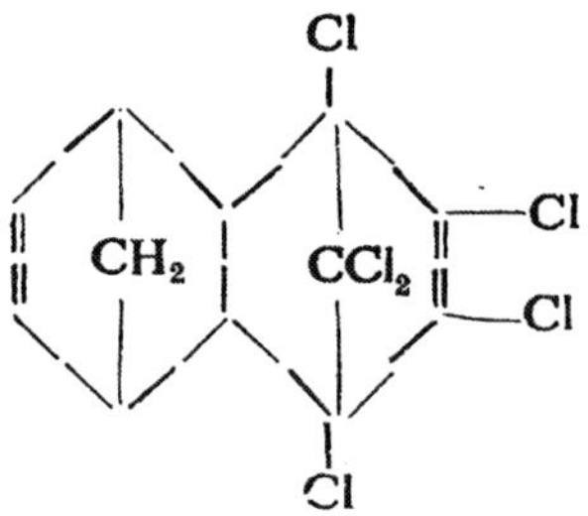	Aldrin	1,2,3,4,10,10-hexachloro-1,4,4α,5,8,8α-hexahydro-1,4-endo-exo-5,8-dimethanonaphthalene
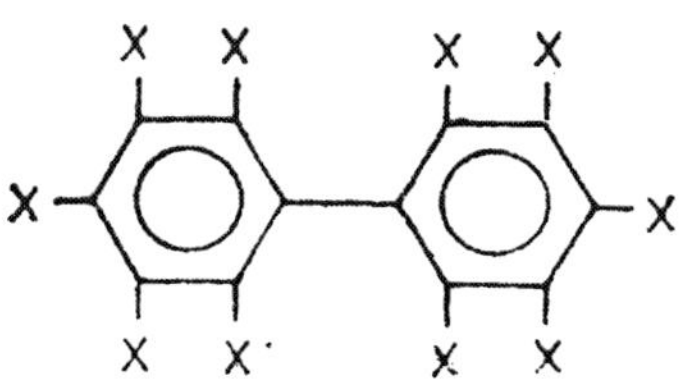	PCB	Polychlorinated biphenyl

x = Possible site of chlorine atom

Fig. 14.1. Structures of Selected Chlorinated Hydrocarbons

to 700 ppm in their bodies. Water fleas can concentrate DDT from 0.5 ppb in the surrounding water to 50 ppm in their bodies. This process of accumulation can continue through food chains until harmful levels are reached in the ultimate predators. The effect is, of course, greater in the case of DDE owing to its greater persistence and prevalence. Accumulation occurs largely because of the affinity of DDT and DDE for fats. This leads to their localization in animals in adipose tissue where the turnover rate is low. Stress may cause mobilization of adipose tissue. The subsequent release of the accumulated residues may lead to toxic levels in the blood of animals even after exposure to the pesticides has ceased.

The harmful effects of chlorinated hydrocarbons were first noted in birds of prey where reproductive failure led to a marked decline in many populations. This was partly due to impaired calcium metabolism which resulted in fragile eggs with abnormally thin shells, and partly due to behavioural changes which favoured egg breakage. This has proved to be a problem in battery chicken farming too, where sawdust from wood treated with chlorinated hydrocarbons has been used as litter. Harmful effects on fish have also been noted, including behavioural changes leading to reproductive difficulties increased mortality among the young and, in some cases, acute toxicity to adults. Further, small concentrations of DDT (0.01 ppm) have been shown to reduce photosynthesis in marine plankton, while even smaller concentrations (1 ppb) in seawater can kill many brine shrimps (*Artemia salina*) within a few weeks.

Agarwal (1983) has given an account of pesticide pollution of waters and has compiled data on mean levels of DDT residue in some worked-out river and lake waters of India. During 1976-78, the DDT residue in River Yamuna at Delhi was 0.249 ppm in upstream and 0.558 ppm in the downstream off Wazirabad.

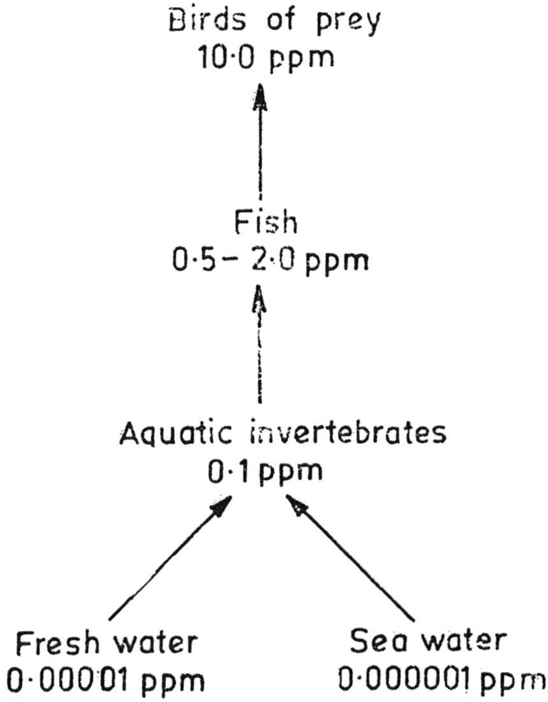

Fig. 14.2. A simplified diagram showing food chain accumulation of DDT.

Another insecticide in this group which has caused environmental damage in the USA is kepone. This compound was used domestically as an ant or cockroach poison, but its production was stopped when it was shown to have caused brain and liver damage, sterility, slurred speech, loss of memory and eye-twitching in workers who had been exposed to it. Matters were made worse by leakage of kepone from the production plant into the main sewer system of Hopewell, Virginia USA and, therefore, into the James River. As a result, fish and shellfish stocks were contaminated and the Governor of Virginia closed more than 100 miles of the river to commercial fishing until such time as residue concentrations should fall to a safe level. Another insecticide, mirex, which is chemically identical to kepone except for one oxygen atom, is used to control ants, but there is concern about its possible carcinogenicity.

Polychlorinated biphenyls (PCBs) are structurally similar to DDT and have similar effects. They are stable at temperatures up to 800°C, resistant to acids, bases and oxidants, and only sparingly soluble in water. Because of these properties, many uses have been found for them. They may act as heat transfer fluids, insulators, hydraulic fluids or plasticizers, and are incorporated in paints, adhesives, gasket sealers, brake linings, fluorescent lamps and carbonless copying paper.

Polychlorinated biphenyls enter the environment accidentally or through poor waste disposal. Like DDT and DDE, they are persistent in the environment because of their chemical stability. Also, their lipid solubility facilitiates food chain amplification. Their acute effects are not a serious problem but their chronic effects are so similar to those of DDT and DDE that it is likely that they act synergistically. Because of this, their use is now carefully controlled in most countries and largely restricted to closed system applications. However, care must still be taken to avoid accidents. This was clearly demonstrated in 1973 when 225—450 kg of the related polybrominated biphenyls (PBBs) were accidentally added instead of magnesium oxide to livestock feed in Michigan, USA. This resulted in contamination of farm animals making necessary the destruction of approximately 29,800 cattle, 5920 pigs, 1470 sheep and 1.5 million chickens. Also rendered unsaleable were more than 880 tonnes of feed, 8.1 tonnes of cheese, 1.2 tonnes of butter, 15.4 tonnes of dry milk products and nearly 5 million eggs. It is impossible to know how many people have suffered sublethal poisoning as a result of eating contaminated meat, eggs and milk, but it may be a very considerable proportion of the population of Michigan. Amongst the possible effects of poisoning with PBBs have been identified teratogenicity and carcinogenicity in rodents, and porphyria and reduced fertility in birds. The origins of this disaster seem to have been due to a shortage of bags prelabelled in red so that the PBBs, marketed as the fire retardant 'Firemaster', were for a time packed in plain brown bags on which the trade names were stencilled in black. Magnesium oxide, a normal feed additive, marketed as 'Nutrimaster' and similar in appearance to Firemaster, was produced in the same plant and packed in identical brown bags, differing only in the stencilled name. Somehow, ten to twenty 50 lb bags of Firemaster were included in a truck-load of Nutrimaster. Because of the superficial similarity of contents and packaging this was not noticed. As a result, animal and poultry feed throughout Michigan was contaminated. To make matters worse, some farmers sold contaminated animals to rendering plants to be processed into livestock feed, dog and cat food as well as tallow and grease used in soaps, shampoos etc. This accident illustrates the importance of clear labelling of toxic substances and of constant vigilance to spot poisoning before it becomes widespread.

Organophosphates

The organophosphate insecticides include malathion, diazinon and parathion. They are readily metabolized by mammalian enzymes and rapidly degraded in the environment. Thus, any environmental damage caused by these compounds will tend to be localized in the area of application. However, these compounds are related to the nerve gases developed for use in war and can be lethal to many different organisms, including humans. Hence, anyone applying these compounds should wear protective clothing, including goggles, rubber gloves and even a respirator. One should avoid treating areas which are important habitats for wildlife.

The principal action of the organophoshate compounds is o1 inhibit acetylcholine esterase. Thus, the action of acetyl choline released at the nerve synapses ceases to be finite with each nerve impulse. Amongst the consequences are tremors in involuntary muscles, convulsions and death.

Carbamates

Among the carbamate pesticides, methyl isocyanate or MIC (CH_3NCO) as an environmental pollutant has drawn attention of public all over the world due to its tragic leakage from the Union Carbide Factory in Bhopal on Dec. 3, 1985 in which over ten thousand persons died and many more were seriously affected. Much of the leaked gas ultimately settled on soil and was washed out to water bodies. Although pesticide pollution of this magnitude is very rare yet smaller incidents are very frequent and together affect the health of about a million persons every year all over the world. The synthesis of MIC requires the use of phosgene ($COCl_2$) which is an extremely poisonous chemical.

Carbamate insecticides, also used as molluscicides, fungicides and herbicides, include carbaryl (sevin), baygon, temik and zectran. They are even less persistent than the organophosphates and less harmful to man. They may cause local environmental problems if used carelessly. Carbaryl, for example, is very toxic to bees. Like the organophosphates, the carbamates act by inhibiting acetyl choline esterase.

Pyrethroids

Pyrethroids are the insecticidal constituents of pyrethrum flowers (*Chrysanthemum cineriaefolium* and *Chrysanthemum coccineum*). Five insecticidal compounds, pyrethrins I and II, cinerins I and II, and jasmolin II, are present in the achenes of the flowers from which they may be extracted with organic solvents. These naturally occurring pyrethroids are unstable to the action of sunlight, air, moisture and alkalis, and are rapidly degraded after application.

Pyrethroids have a low toxicity to mammals because they are rapidly metabolized to harmless substances. However, they can cause severe allergic dermatitis and systemic allergic reactions. Large amounts may cause nausea, vomiting, headache and other disturbances of the central nervous system.

The high cost of extraction of natural pyrethroids has led to the production of similar synthetic compounds, of which the best known are the allethrins. The synthetic compounds have a wide range of properties and it is impossible to generalize about them.

Nematicides

Against Plant Parasitic Nematodes

The nematicides can affect large areas of land. The main nematicides used are 1, 3-dichloropropene (telone) and 1, 2-dibromoethane, both of which are liquids applied directly to the soil. Because they are phytotoxic they can only be used in the absence of growing crops. These compounds are also very toxic to mammals. They irritate the skin, eyes and mucous membranes, damage the liver, and ultimately cause death from severe lung injury. Other soil nematicides in use are sodium N-methyldithiocarbamate in aqueous solution O-2, 4-dichlorphenyl-O-diethylphosphorothioate (V-C 13 nemacide) as an emulsion, calcium cyanamide, urea, and 3, 5-dimethyl-tetrahydro-1, 3, 5, 2H-thiadiazine-2-thione (mylone). The last three chemicals are solids and must be thoroughly mixed with the soil to be effective. All of these compounds are toxic to mammals and probably to most other animals as well.

Fumigants, such as methyl bromide, may be used in confined spaces, or under polythene sheeting, where only small areas require to be treated. Methyl bromide is very toxic and exposure of mammals to 50 mg per body weight can cause death.

Seeds and bulbs may sometimes be treated directly with

dilute formaldehyde solution in water (0.5 to 1.0% weight to volume), with or without a trace of detergent. It is unlikely that this constitutes an environmental hazard.

Against Mammalian Parasitic Nematodes

Whether the drugs used in the treatment of nematode infection in human beings and domesticated animals constitute an environmental hazard does not seem to have been assessed. However, since at least 200 million people throughout the world are thought to be infected with nematodes, and since presumably a similar proportion of the domesticated animal population is at risk, the use of nematicidal drugs must be considerable. In consequence, appreciable amounts of these compounds or their derivatives must enter the environment. It is, therefore, justifiable to describe briefly here the principal drugs involved.

Diethylcarbamazine is the most effective drug against filarial nematodes. It is not highly toxic but it can cause nausea, vomiting and headache in humans, and muscle tremors and convulsions in dogs. Thiabendazole is the most widely used drug for the control of gastro-intestinal nematodes, other than *Trichuris* spp., in ruminants. It is also used on citrus fruit as a fungicide. Its acute toxicity to mammals is low. Phenothiazine is often used against intestinal nematodes in sheep and chickens. It is not used on cattle, horses or humans, because of its many toxic effects on these species. Piperazine is the drug of choice for the treatment of oxyuriasis and ascariasis in both humans and other mammals, since it is highly active against the nematodes responsible but almost nontoxic to mammals at the therapeutic dose levels.

Fungicides

Probably the most widely used agricultural fungicides are Bordeaux mixture and sulphur. Bordeaux mixture is made by combining a solution of copper sulphate with a suspension of calcium hydroxide, usually in equal amounts by weight. Bordeaux mixture may have harmful side effects on treated plants, e.g., it may cause accelerated transpiration and, hence, skin russeting of apples, and premature defoliation of peach trees. Other effects may be associated with localized concentrations of copper. Sulphur has low toxicity but it may cause irritation to the skin, eyes and respiratory tract. It may also be converted to hydrogen sulphide which is a very powerful poison.

Organic fungicides are replacing Bordeaux mixture and sulphur in many applications. Organomercurial compounds are used as seed dressings. Dithiocarbamates such as ferbam, ziram, nabam, zineb and maneb, are widely applied to vegetables and ornamental plants. Other organic fungicides in use are thiram, captan, glyodin, chloranil, dichlone, dodine (cyprex) and karathane. Of these, thiram, dodine and karathane are appreciably toxic to humans. Captan, glyodin, chloranil and dichlone have low toxicity for humans because they are poorly absorbed from the gut.

Hexachlorobenzene, which has been used as a fungicidal seed dressing, has a relatively low acute toxicity for mammals but can cause serious delayed effects. This was seen most dramatically in Turkey in 1955 when up to 5000 people were poisoned by eating hexachlorobenzene-treated grain. Their symptoms included enlarged livers, abnormal light sensitivity, weight loss and abnormal hair growth, particularly on the face. This last symptom gave the epidemic its name of monkey face disease. It is worth noting that hexachlorobenzene may be an impurity in other pesticides, at levels up to 10 per cent of the total. It is chemically stable and, therefore, persistent. Measurable amounts are released into the environment during the manufacture of perchlorethylene.

The antibiotic, cycloheximide, has been used to treat cherry leaf spot. This antibiotic is extremely toxic, causing gastrointestinal lesions, coma and death. Sympathomimetic drugs may alleviate the symptoms and hydrocortisone or adrenal cortical extract may prevent death.

In addition to the fungicides which are applied directly to plants or seeds, there are others used to clear fungi from the soil before planting crops. Many of these, *e.g.*, formaldehyde, chloropicrin and methyl isothiocyanate, are phytotoxic but their volatility enables them to escape from the soil before planting commences. However, these compounds are also highly toxic to animals and this should be borne in mind when their application is being planned. Pentachloronitrobenzene is a soil fungicide which appears to have low phytotoxicity and it has been applied directly to soil in which cruciferous crops and lettuce are growing. However, it is moderately toxic to animals and accumulates in body fat. Hence, there is a potential long-term risk similar to that from chlorinated hydrocarbons.

Herbicides

The toxicity of herbicides to animals is low, but it is rarely negligible. Repeated doses of diuron have been shown to produce anaemia and methaemoglobinaemia in rats. A number of people have died following accidental ingestion of paraquat. Others have died following excessive exposure to derivatives of 2, 4-dinitrophenol. Even the widely used chlorophenoxy acid herbicides can cause eye irritation and gastro-intestinal disturbances in animals exposed to large amounts. One of the chlorophenoxy acid herbicides, 2, 4, 5-T may contain appreciable quantities of dioxin as an impurity. Dioxin is the name given to the family of chemicals related to 2,4,7,8 tetrachlorodibenzo-p-dioxin (TCDD). This family has been shown to be strongly teratogenic though its most obvious effect is a skin disease called chloracne. It is also environmentally stable and may accumulate. The EPA has now banned most crop related uses of 2, 4, 5-T, its use on cattle pasture has been questioned and animals must not be grazed on treated areas within two weeks of slaughter.

Inorganic Pesticides and Herbicides

In the past, various compounds of zinc, copper, arsenic. mercury and other metals have been used as pesticides.

Sulphuric acid, sodium arsenite, sodium chlorate, sodium penta and metaborate and sodium thiocyanate have been used as herbicides. Some of the herbicides, e.g., sodium chlorate, are still used for total ground clearance but, in general, the use of inorganic pesticides and herbicides is much less than it was. This is a welcome development because all these inorganic compounds are highly toxic to a wide range of organisms and the metallic derivatives may leave persistent toxic residues in the soil. Nevertheless, it may still be necessary to use such compounds where pests have developed resistance to the more selective organic pesticides.

REFERENCES

Carter, L.J. 1976. Michigan's PBB incident: Chemical mix-up leads to disaster. *Science*, **192**, 240-243.

De Bach, Paul. Ed. 1964. *Biological Control of Insects Pests and Weeds.* Reinhold Publishing Corporation. New York.

Gunn, D.L. and J.C.R. Stevens, (Eds). 1976 Pesticides and Human Welfare. Oxford University Press.

Higgins, I.J. and R.G. Burns, 1975. *The Chemistry and Microbiology of Pollution.* Academic Press, London.

Varo, P. (Ed). 1976. *Pesticide Chemistry*—3 3rd IUPAC Congress on Pesticide Chemistry- Butterworths. London.

15

Oil Spills and Radioactivity

Petroleum and Related Compounds

Petroleum may be toxicant in the sense that it causes harm to living organisms as a result of some chemical action. Much of the environmental damage it causes is due to its physical properties. The most obvious effects are due to its hydrophobic nature and its impermeability to oxygen. Thus, when it coats organisms, they die from asphyxiation (lack of oxygen). Similarly, contaminated areas of the environment suffer from oxygen depletion. The hydrophobic properties are associated with chemical stability. Therefore, environmental damage may occur long after imput to the environment has ceased. For this reason special precautions must be taken to minimize the anthropogenic contribution.

Petroleum

Petroleum, generally referred to simply as oil, is a mixture of alkanes containing, on average, about 1% by weight of aromatic hydrocarbons. It has become a matter of serious environmental concern because its extraction and use as an energy source by humans has led to its widespread distribution in the biosphere. This has been most obvious following large-scale marine spillage, either from tanker wrecks or from oil-well blowouts, but the cumulative effects of the many minute spills that have occurred both on land and sea may be at least as serious.

The harmful effects of oil on living organisms may be divided into two (i) that are primarily physical and (ii) that are primarily chemical. Physical effects are caused by oil coating the organisms or their immediate environment. This is very cleary seen when water birds become covered with oil. By matting the feathers, the oil destroys their insulative capacity, reduces buoyancy in the water and prevents flight. In other organisms, oil coating may cause death by asphyxiation. Oil films on the surface of natural water reduce light transmission and, hence, photosynthetic primary production. Such films also retard oxygen uptake by water and so cause a lower dissolved oxygen concentration and the death of many organisms.

Chemical effects of oil can be related to the components involved. Low boiling point saturated hydrocarbons, at least up to octane, can produce anaesthesia and narcosis in many lower animals. Low boiling point aromatic hydrocarbons are even more toxic, and their greater water solubility tends to enhance their distribution and uptake by aquatic organisms. Benzene, toluene, naphthalene and phenanthrene are amongst the compounds in this group. Benzene characteristically inhibits blood cell formation in bone marrow. All cause local irritation of the respiratory system and excitation or depression of the central nervous system. Many may be mutagenic, carcinogenic or teratogenic. Polycyclic aromatic hydrocarbons are especially dangerous in this respect. High boiling point saturated and aromatic hydrocarbons may not exert much direct toxicity but may interfere with the responses of aquatic organisms to chemical stimuli, such as, sex attractants, with equally serious consequences.

Since many of the components of oil are chemically stable and not readily metabolized or excreted once absorbed, they are subject to food chain amplification. In this way, they may give unpleasant flavours to food for human consumption or render it toxic. It is even conceivable that oil in natural waters may concentrate fat soluble toxicants by a partitioning process, thus enhancing their toxicity.

Not only are there hazards associated directly with oil spills, there are others associated with oil clearance from natural waters and from beaches. Containment with booms followed by removal by skimming devices causes the least damage but is frequently impossible. As an alternative, detergents may be used to disperse the oil. This may give aesthetically satisfactory results but may enhance the biological destruction done by the spill. All detergents have appreciable toxicity and can facilitate uptake of oil by aquatic organisms. Further, they may cause oil on beaches to penetrate the sand more deeply, thereby prolonging harmful effects on intertidal animals and plants.

Polycyclic (Polynuclear) Aromatic Hydrocarbons

These compounds, e.g. 3,4-benzpyrene and 1,2-benzanthracene have attracted considerable attention. They are believed to be the main carcinogens in cigarette smoke and are potent mutagens. They are found in petroleum and in industrial products, and are also synthesized by plants. They cause mutations either by direct covalent bonding with DNA or by wedging themselves into the DNA helix (intercalation).

Much more needs to be known about the natural background level of these compounds and the extent to which organisms have evolved metabolic systems, e.g., the cytochrome P_{450} system to detoxify them. Even if such systems are available, care must be taken not to overload them. It would, therefore, seem prudent to minimize the anthropogenic production of these compounds.

Radioactivity

Radioactivity is toxicant in the sense that it causes harm to living organisms. Most harm is caused by toxizing radiation.

This radiation modifies or inactivates biological molecules, frequently with fatal results.

Radionuclides

Radionuclides are elements which spontaneously disintegrate (decay) into smaller particles and emit ionizing radiation. Certain types of this radiation, e.g., alpha particles and neutrons, have a high energy transfer per unit path length. This means that they do not travel far from the source, i.e., no more than 70 μm in soft tissue, but have pronounced effects on any atom or molecule with which they come in contact. Even at low doses the effects can be very serious, but the risk is highly localized. Beta particles (electrons) can travel farther, i.e., up to a few mm, but are still not very penetrating. Their effects at low doses can also be serious. On the other hand, high energy X-or gamma-rays are very penetrating but, because of their low energy transfer per unit path length, tend to have little effect at low doses, e.g., the 27 to 30 millirads usual in medical chest X-ray. The rad (radiation absorbed dose), is defined as 0.01 J kg^{-1}, i.e., it is a measure of the amount of radiation energy absorbed per unit mass of material. However, one rad of alpha-radiation is about four times as damaging as one rad of X-rays. Therefore, in order to compare doses in terms of biological effects, another unit is used, the rem. One rem is biologically equivalent to one rad of X-rays. As the rad and rem are very large units relative to normally occurring radiation, most of this is measured in millirads or millirems. Recently, SI units for radiation have been introduced. In these, the gray (Gy) replaces the rad, and the sievert (Sv) replaces the rem. The relationship between these units is shown in Table 15.1.

The disintegration rate of a radionuclide is defined in terms of its half life, which is the time taken for half the atoms in a given sample to disintegrate. Radionuclides with short half lives, i.e., up to a few days, may be very dangerous when produced, but the danger does not last. Those with very long half lives, i.e., more than 10^5 years, are usually of such low activity as to be fairly safe. Therefore, those with intermediate half lives have the greatest environmental significance.

TABLE 15.1. Units for Measurement of Radioactivity

Physical quantity	*SI unit*	*Non-SI unit*	*Relationship*
Activity, i.e., nuclear transformations	becquerel (Bq) 1 Bq = 1 s^{-1}	curie (Ci)	1 Bq = 2.7×10^{-14} Ci 1 Ci = 3.7×10^{10} Bq
Absorbed dose	gray (Gy) 1 Gy = 1 J kg^{-1}	rad	1 Gy = 100 rads 1 rad = 0.01 Gy
Dose equivalant	sievert (Sv) 1 Sv = 1 J kg^{-1}	rem	1 Sv = 100 rems 1 rem = 0.01 Sv

Radionuclides occur naturally in the environment. Radionuclides are also produced directly or indirectly from human use of uranium. As a result of mining and processing ores to produce usable radioactive materials, large piles of waste 'uranium tailings' are left behind and radionuclides can be leached from these by rainfall, etc. Atmospheric explosions of nuclear weapons produce fallout of radionuclides over large areas of the earth's surface. Two components of this fallout have caused considerable concern, i.e., strontium-90 and caesium-137. In nuclear power plants radionuclides are formed when impurities in the primary coolant water are bombarded with neutrons from the fuel elements in the core. Even if the coolant water is free of impurities radionuclides may escape into it from the steel or zirconium containers holding the radioactive fuel. Thus, the coolant water must be constantly monitored so that release of radionuclides into the environment can be minimized. With time, radionuclides produced by nuclear fission accumulate in the nuclear reactor fuel and slow or stop the fission reactions. At this point, the fuel elements must be removed and processed to extract the fission products. The disposal of these nuclear wastes presents considerable problems. Liquid and slurry wastes from reprocessing are stored in steel or concrete tanks until the radioactivity falls to acceptable levels for discharge and dispersal, often into the sea, but, as some of the nuclides are very long-lived, there could be a long-term problem caused by their accumulation, possibly far from the point of discharge. Solid materials, with low levels of radioactivity, may be mixed with cement or other immobilizing agents and sealed in steel drums for burial, but some radioactivity may still remain when the drums eventually corrode. High level materials present severe problems caused by the high temperatures resulting from the radioactivity, by the longevity of some of the products (half lives of thousands of years) and by their toxicity. The most promising method of disposing of them seems to be vitrification, i.e., conversion to a glass-like product, followed by burial at a great depth under the earth's surface. Very low level discharges of radionuclides may be made directly into sewers or into the atmosphere from high chimneys.

Effects

When ionizing radiation from a radionuclide hits an atom it strips off an electron and releases a large amount of energy. Thus, almost any biological molecule hit directly by ionizing radiation is functionally destroyed. This, in itself, is enough to cause considerable damage to a living cell but the damage is accentuated by the formation of highly reactive free radicals. Since water is the main component of any cell, most of these radicals are either H or OH. Radiation damage of any kind is greatest in the presence of oxygen.

Direct damage to DNA, or indirect damage from reaction with free radicals, causes chromosome breaks, cross-linking of the polynucleotide chains or modification of the constituent bases. Various repair processes can occur but there is still a high probability of mutations being produced. Most of these mutations are lethal and the rest are frequently teratogenic or carcinogenic.

Damage to proteins is associated with a general drop in enzyme activity which has been attributed mainly to the splitting of disulphide bonds. Hydrolytic enzymes may appear to become more active, but this almost certainly reflects their release from lysosomes, following damage to lysosomal membranes. Once released, they rapidly destroy the cell.

The net effect of ionizing radiation is to increase mortality. This follows from harmful mutations, reduced metabolism and, in higher animals, from lowered resistance to disease following depression of the immunological system.

Strontium-90 and caesium-137, in fallout from nuclear weapons, have attracted considerable attention because of their ready accumulation by living organisms. Strontium-90 is similar to calcium and becomes localized in bones, where the emitted radiation can damage the blood cell forming bone marrow and cause anaemia or leukaemia. Caesium-137 is similar to potassium and is actively absorbed by all cells. Thus, the chances of radiation damage are greatly enhanced.

Of the high level radiation products of nuclear reactors, plutonium-239 has caused the most concern. It emits α-radiation. Like strontium-90 it accumulates in bone, with similar consequences of anaemia and leukaemia. Combustion of plutonium produces plutonium dioxide particles, any one of which, if inhaled, has a high probability of causing lung cancer.

REFERENCES

Beynon, L.R. and E.B. Cowell, 1974. *Ecological Aspects of Toxicity Testing of Oils and Dispersants* Applied Science Publishers, Barking.

International Atomic Energy Agency, 1963. *A Basic Toxicity & Classification of Radio-nuclides.* Technical Reports Series No. 15, International Atomic Energy Agency, Vienna.

16

Noise Pollution

Sound and Noise

Noise is any sound that is not wanted. It is one of the more common forms of atmospheric pollution. Because individual perception and interpretation of sounds differ, what is music to one person may be noise to another. Family disagreements over phonograph and radio selections are very familiar. The city dweller who visits his relatives in the country cannot go to sleep in the evening because it is too quiet; and he loses sleep in the morning because the rooster crows and the dogs bark. By contrast, the suburban or country dweller is apt to experience sleepless nights in the city due to the annoyance of loud music in a neighbour's apartment and traffic noises. Daytime noises are even more intense.

Injurious Effects of Noise

Violent noises can cause temporary or permanent impairment of hearing, thus the expression "deafening." Continual noise can lead to gradual decline in auditory acuity and eventual deafness. Prolonged exposure to noises may have other effects that are more subtle but possibly as serious.

Noise causes several undesirable effects. Damage to hearing

and loss of sleep are only two of the more obvious insults to the relationship between man and his environment. Noise interferes with speech, sometimes making it unintelligible; sounds of warning are misunderstood or not heard. Even low-level noises impair verbal communication because they require more attention and effort and cause misinterpretations. Noises produce irritability and a feeling of fatigue and may reduce a worker's efficiency. There is evidence that noise is one of the major causes of stress and many of the other human afflictions associated with tension—anxiety, insomnia, accident-proneness, high blood pressure, and other cardiovascular diseases. Even in its less serious forms, noise is commonly a source of annoyance and cause for complaint. All these are of personal, social, and economic importance.

The effects of noise on the foetus are not fully known. Medical scientists have noted that an unborn child will move and kick when there is a loud noise. It also responds with a sudden increase in the heartbeat as though it were disturbed or frightened.

Nature of Sound and Noise

Sound is produced when an object vibrates, alternately compressing and expanding the air. The compressions and expansions travel wavelike from the source. They are called *sound waves* or simply vibrations. The two most basic characteristics of sound are *pitch* and *amplitude*, and these are related to *loudness*.

Pitch

When an object serves as a source of sound, the pitch is determined by how rapidly the object vibrates. The rate of vibration is called the *frequency*. The higher the frequency, the higher the pitch, and the lower the frequency, the lower the pitch. This can be easily demonstrated by playing a phonograph record at a slower speed than intended. The words or tones will not only emerge more slowly, but will have a throaty sound and a distinctly lower pitch. The reverse effect is produced by playing the record at an excessively high speed. Similar results are produced as the speed of an electric fan or that of a siren increases.

Technically, the *frequency* of an object, or of its sound waves, is defined as the number of vibrations per second. People can hear sounds with frequencies from about 20 to 20,000 vibrations per second, but there is a great deal of individual variation. Some persons can hear frequencies that others are not able to detect, and sometimes there is poor hearing in part of the frequency range due to disease or injury.

The notes on a piano range from a frequency of 27 vibrations per second for the lowest to about 4,000 for the highest. The quality of a stereo system is determined partly by the range of frequencies that it can reproduce with a high degree of integrity. There are also differences between humans and animals in the ability to detect different frequencies. Bats produce and hear sounds having frequencies far above the range of the human ear. Nature is full of sounds to which humans are insensitive.

A change in pitch occurs as one moves rapidly toward or away from a source of sound, as when a train-rider approaches or leaves a railway crossing signal. The lower pitch caused by "stretching out" the sound waves, or the reverse effect produced by "squeezing together" the sound waves, is called the *Doppler effect.*

Amplitude

Sounds having the same frequency, and therefore the same pitch, may differ in the amount of energy involved. The energy that goes into producing a sound determines the intensity of the sound. But intensity and loudness differ; loudness is the strength of the *sensation* of sound perceived by the individual.

The distance that a vibrating object, such as a tuning fork or a guitar string, moves as it vibrates is called the *amplitude of vibration.* The greater the energy that goes into producing the sound, the greater the amplitude of the sound that is produced. Also, a large vibrating object will produce a sound of greater intensity and loudness than a small vibrating object. This can be demonstrated by making the prongs of a table fork vibrate, then placing the handle of the fork on the top of the table. The sound becomes instantly louder because the table top increases the area of vibration. The intensity and loudness of sound are also affected by the density of the medium. The sound produced by striking two rocks together under water is louder than when they are struck together in air because the denser the material, the louder the sound. There is no sound in a vacuum because there is nothing there to form sound waves.

Loudness

The loudness of sound (the strength of the sensation) does not depend entirely on the energy or amplitude of the sound wave. Loudness also depends on the frequency or pitch. The human ear has a low sensitivity for sounds of low pitch and for those of extremely high pitch. For example, to produce the same sensation of loudness, a low pitch that has 64 vibrations per second must be 10,000 times as powerful as a high pitch of 3,000 vibrations per second. The maximum sensitivity comes at about 1,000 to 3,000 vibrations per second. (The frequency range of speech is from about 200 to 1,000, while middle C on the piano has a frequency of about 256).

For any given frequency, the energy of a sound must be increased about 100 times to make the sound 2 times as loud, and must be increased about a million times to make the sound 6 times as loud. If we notice that $100=10^2$ and $1{,}000{,}000=10^6$, it can be seen that loudness increases only by the exponent of 10 (2 and 6) and not by the energy required. The unit of loudness expressed by the exponent of 10 is called a ***bel.*** In the first example the loudness is 2 bels and in the second example the loudness is 6 bels.

The term *decibel* is often used to indicate one-tenth bel.* A sound of between 0 and 1 decibel is about the weakest that the average human can hear. For testing purposes, 0 decibel is considered to be the threshold of hearing. A whisper is about 20 to 30 decibels and an average speaking voice is about 60 decibels. Note that according to the logarithmic relationship, 20 decibels is 100 times as intense as the threshold and 60 decibels is 1,000,000 times as intense. The loudest sound that a person can stand without discomfort is about 80 decibels (Table 16.1). Automobile horns may reach 90 decibels and a jet airplane at a distance of 30m may have an intensity of about 140 decibels.

TABLE 16.1. **Examples of Noise Levels and Human Response**

Source of Noise	*Decibels*	*Human Response*
Jet takeoff (nearby)	150	Threshold of pain
Hydraulic press (I metre)	130	Limit of amplified speech
Jet takeoff (60 metres)	120	Maximum vocal effort possible
Construction noise (3 metres)	110	Very annoying
Heavy truck (15 metres)	90	Hearing damage (8 hours)
Alarm clock	80	Annoying
Highway traffic (15 metres)	70	Telephone use difficult
Air conditioning unit (6 metres)	60	Intrusive
Quiet residential area	50	Quiet
Living room	40	
Soft music	30	Very quiet
Broadcasting studio	20	
Rustling leaves in breeze	10	Just audible
	0	Threshold of hearing

*Technically, a decibel is the amount of sound wave pressure that equals 0.0002 dyne per square centimetre. A dyne is a force which if exerted for 1 second will move 1 gram a distance of 1 centimetre.

Kinds of Sound

The kind of sound—whether noisy, musical, abrasive, soothing, displeasing, or pleasant—depends on a combination of characteristics in addition to pitch, amplitude, and loudness. A vibrating object produces many frequencies even though, as with a musical instrument, one tone may predominate. A blend of many frequencies produces a characteristic sound, which in music is called *quality.* It distinguishes a guitar, for example, from a mandolin. The vibrations that occur in addition to the fundamental frequency are called *overtones* or *harmonics.*

If two objects vibrate at different frequencies, the result is pulsations of loudness called *beats.* Since the number of beats per second is equal to the difference in frequencies of the two objects, the closer the tones, the faster the beats. When the number of beats reaches about 30 per second from tones that are very close but not quite the same, the sound is very unpleasant. Musicians call this a form of *dissonance.* At beats above 30 per second, the frequencies are so close that the human ear cannot detect the beats. Since the ear cannot tell that anything is wrong, the disagreeable sensation disappears.

Whether sound is pleasant or unpleasant is determined partly by *resonance,* produced when repeated vibrations of the same frequency create a strong response. A sound box of the right size and shape increases resonance. *Sympathetic vibrations* are a form of a resonance produced in an unattached object having the same natural resonance as the object that produces the sound. For example, dishes in a cupboard or a picture on a wall may rattle when a certain note is sounded on the piano. A comparable situation is when a strong wind creates resonant vibrations in a suspension bridge. The bridge may fall if the vibration becomes intense. Soldiers are always ordered to break step when crossing a bridge because when marching in unison, they can set up vibrations that may cause the bridge to break apart.

Speed of Sound

Sound travels at different speeds, depending on the density and elasticity of the medium through which it travels. The denser and less elastic the substance, the slower the speed of the sound waves. Sound travels at a velocity of 344 m s^{-1} in air when at a temperature of 20°C. When the temperature drops to 0°C, the air is denser and the sound waves travel at a slightly slower speed. The velocity increases about 0.6 m s^{-1} for each centigrade degree rise in temperature. Water is denser than air but also more elastic. Sound travels through water at a velocity of about 1,440 m s^{-1}. Steel is 6,000 times denser than air and this quality retards the movement of sound waves, but steel is also much more elastic than air by a magnitude of 2 million times; therefore, sound travels more rapidly through steel, at a velocity of 5,000 m s^{-1}.

Thus the speed of sound, regardless of the substance through which it travels, is much slower than the speed of light, which is 3×10^8 m s^{-1}. It is easy to estimate the distance from the source of a far-off sound if there is some other signal to identify it, such as a puff of smoke or a flash of light. Five seconds between a lightning flash and a clap of thunder equals a distance of about 1.6 km.

Sources of Noise

The sources of noise pollution in an industrial society are manifold. The main source in cities and other highly developed areas are noises associated with transportation. Automobiles, trucks, motorcycles, airplanes, freight trains, and accessory noisemakers such as horns, sirens, and other devices produce a variety of noise that has few equals on earth for intensity and annoyance. Much of it is unnecessary. The use of horns and sirens, except in extreme emergencies, is undesirable. Cities that have passed laws against horn-tooting have reported no increases in accidents.

Noises associated with manufacturing, building construction, are second in intensity to transportation noises. Because noise suppression is often costly, voluntary corrective measures cannot be depended on. Legislation to enforce reasonable standards would probably be needed to satisfactorily alleviate the problem.

Aircraft, due to their increase in size and enormous increase in numbers, have become one of the most troublesome sources of noise pollution. The scream of jet engines is not conducive to tranquility and repose. In some locations, conversations stop every few seconds, television viewing and listening becomes impossible, and sleep and relaxation are interrupted. This condition is multiplied many times throughout the country and the world.

Control of aircraft noise requires several changes. Procedures already in use call for prescribed courses and reduction of power at certain altitudes. Other methods being tested include a study of climb-out profiles, glide slopes for landing, and the design and development of quieter engines and airplanes. A basic problem is that the safest jet approach to a landing is a long, low approach; but this is also the noisiest. One solution is to rezone the area under the landing and takeoff pattern, but this is costly and seldom solves the problem. Relocation of airports is another solution.

The supersonic transport (SST) brought noise pollution problems. A supersonic airplane is one that travels faster than sound. When an aircraft travels at a velocity greater than the speed of sound, there is an enormous increase in air resistance. A *shock wave* is created that can be extremely energetic, depending on the size, speed, and route of the aircraft. When a shock wave strikes things below, it can have a jolting and devastating effect. The shock wave, or *sonic boom*, may spread over an area of 15 to 125 km. Although many cases of damage to structures caused by sonic booms from relatively small military aircraft have been reported and numerous damage claims have been filed, the destructiveness and nuisance effects of large commercial supersonic aircraft have not been fully evaluated. Objections to the development and commercialization of an SST are based on the possibility that their shock waves might become intolerable.

Noise Abatement

In some industries, noise control does not necessarily mean noise reduction. Earplugs worn by shipyard workers may bring about satisfactory control without changing the noise intensity. One way to reduce airport noise is to limit activities to daytime and early-evening flights. Although there is no reduction in total noise, the annoyance level is improved. Another way to cut down noise is by the construction of barriers or insulation; this does not put an end to the noise, but results in the harmless absorption of a portion of the sound waves.

Thus, there are four basic ways to reduce noise and its damaging effects. The first and most satisfactory way is to *reduce the noise level* at its source, for example, by designing and developing quieter machines, trucks, airplane engines, automobiles, and household appliances. The second method is to use *insulation* or absorbing materials, such as the installation of acoustic tile and the construction of enclosures around industrial machinery. The third method may be described as *operational*, such as rescheduling airline flights. Finally, the fourth, and least satisfactory method from the long-range view, is to *protect personnel* by the use of such devices as earplugs.

A quiet atmosphere in which to relax or work is rapidly becoming one of the most urgent environmental requirements. The solution must depend on a combination of methods. However, more emphasis is needed on reducing the noise at its source.

17

Soil and Edaphic Environment

Soil

The nature and quantity of vegetation that can be grown on a given piece of land depend strongly on the characteristics of the soil. Study of the properties and distribution of soil types is therefore essential both for an understanding of existing patterns of vegetation and for determining strategies and prospects for success in increasing the yield of plant materials used by human society.

Soil is the weathered top surface of earth's crust constituted by mineral matter (sand, silt and clay), organic matter in different stages of decomposition (humus), microorganisms (bacteria, fungi, etc.) mixed together in such a way as to have capillary and non-capillary pore spaces filled with moisture and air. This is the medium of anchorage and supply of nutrients and water for the plants and in turn the plants are the ultimate source of food for all animals and man. Thus soil is the basic and most important life support component of biosphere.

Soil functions as a source and as a storage reservoir of water and mineral nutrients for plants, as a medium in which chemical transformations influencing the availability of those nutrients takes place, and as an anchor for the support of terrestrial as well as many aquatic plants. In composition, it is a mixture of inorganic minerals, organic matter, soil animals and microorganisms, gases, and moisture.

Normally, we do not feel concerned about soil as we tend to think that it is most abundant and a never exhausting resource found all over the land surface. A city dweller may even regard it only as a nuisance in the form of dust. But with the rise in human population requiring more and more production of food, the importance of soil and its fertility has now been well recognized. Even in ancient times the importance of soil as a resource for crop cultivation was recognized and human settlements established on plain cultivable lands where good soil and abundant water were available round the year. Man has also realised that all living beings are ultimately the product of soil, and after death, the body materials are returned to the soil.

More than 3000 years ago Mosses had said..."dust thou art and unto dust shalt thou return" emphasizing the fact that out of dust the food is produced which goes to build our body and after death, the body slowly decomposes and becomes part of soil. The cycle may be long for the nutrient elements. From the absorbed surface of colloidal clay and humans it enters the roots, gets transported to leaf, enters biosynthetic pathways to become a complex organic compound that goes to build the body of the plant or some organelle. Then some of it is grazed by a herbivore, and some others remain stored in stem for a long period and still some others are returned to the soil in the form of leaf and twig litter. Some of it goes into the formation of seeds and propagules. Herbivores, carnivores and decomposers ultimately derive their food from one of these sources. Thus soil is the foundation of biosphere and the fountain head of life on the earth. Soil is a product of climatic, physiographic and biotic forces acting upon the rock surface in geological and recent past. The process of soil formation is called *pedogenesis* and the forces that bring it about are collectively called as *weathering*. One very important fact that must be borne in mind is that soil is a living system maintained at its position by the binding and protective effects provided by plants. Its nutrient or fertility status is delicately balanced by the input of dead organic matter and the decomposer micro-organisms that harbour soils. A soil devoid of its microfauna and microflora or of organic matter is no good. A soil, if kept devoid of plant growth for a few years is bound to get eroded and lost. While it takes several decades or hundreds of years to build a thin column of soil, it would take only a few years of misuse to get it lost for ever. Eroded soils with exposed rocks cannot support life. Deforestation leads to erosion and erosion leads to desertification.

Soil Origins and Structure

The main source of the inorganic particles in soil is the weathering of the various types of rock—igneous, sedimentary, metamorphic—by exposure near Earth's surface to temperature changes; the physical action of ice, water, and roots; and chemical attack by rainwater, surface water, and atmospheric gases. Uplifting of unconsolidated sediments from the seafloor is another; rarer source of soil materials. These sources of the mineral components of soil are called *parent materials*; the parent materials can be either *residual* (meaning the soil is formed where the weathering takes place) or *trasported* (meaning the materials have been moved to the place of soil formation by the action of wind, water, ice, gravity, or a combination of these.

Alluvial is the term given to soils deposited by river flow. *Colluvial* refers to soils whose parent materials were deposited by landslides.

Most soils are not uniform in vertical profile but rather are characterized by zones called *horizons*. The upper-most layer or zone is called the *A horizon*. It is the home of most of the soil organisms and the location of the greatest abundance of roots. It has usually been somewhat depleted of soluble substances by leaching. The next zone proceeding downward is the *B horizon.* It receives downward-moving minerals leached from above and, often, upward-migrating substances from weathered parent materials below; therefore it is called the zone of accumulation. The *C horizon* consists of the weathered rock material. The bedrock below, usually but not always the true parent material, is the *D horizon.* (Fig. 17.1).

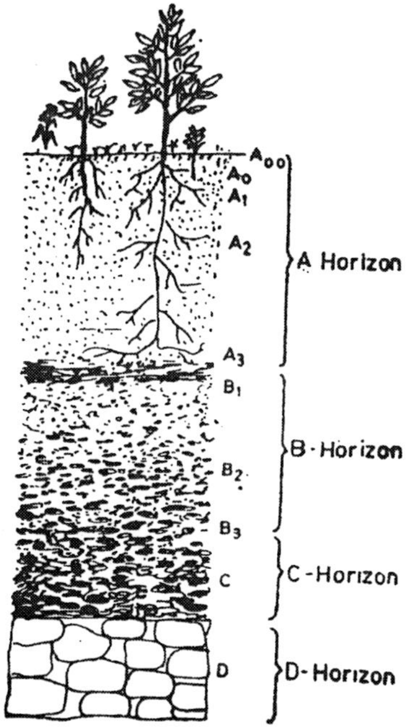

Fig. 17.1. Soil profile showing different zonations.

In soil studies more detailed than our treatment here, the horizons are often subdivided by the addition of subscripts running 1 to 3, top to bottom, (that is, A_1, A_2, A_3). In that notation, the fresh dead organic matter (called *litter*) is the A_{00} horizon; and the partially decomposed organic matter (called *duff*), just above the A_1, horizon, is the A_0 horizon.

Soils in which the layered structure is well developed and distinct are called *zoned normal*; those without this well-developed vertical profile are called *azonal.* Alluvial and colluvial soil are often azonal.

Texture: Another important characteristic of soil is *texture* (meaning the size of the individual mineral particles). An international classification system defines particles with diameter less than 0.002 millimetres as *clay*, those with diametres between 0.002 and 0.02 millimetres as *silt*, and particles between 0.02 and 2 millimetres in diameter as *sand*. The term *loam* is used to describe mixtures of the different size classes of soil particles. A standard classification scheme based on the proportions of clay, silt, and sand in soil is illustrated in Fig. 17.2.

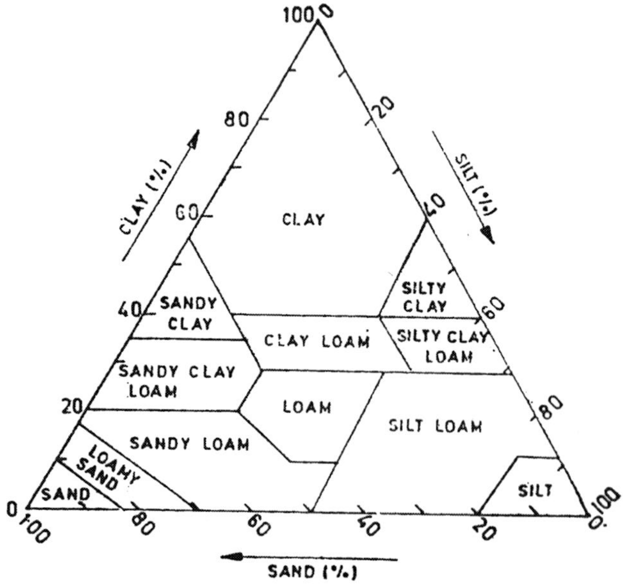

Fig. 17.2. Soil classification by texture. (From Janick et al.)

Soil texture influences the rate at which water percolates through soil and the amount of water a soil can hold. Coarse soils are characterized by rapid infiltration, and hence low surface runoff, but they cannot retain much water. The fine-textured clays are penetrated by water only slowly but have a high storage capacity. The *pore space* in soil, which is filled in varying proportions by water and air, is typically around 50 per cent in many kinds of soils; what is more important than the total volume of the pore space is the characteristic size of individual pore spaces. Few large pores make a much less satisfactory soil than many small ones. Soil organisms and organic matter are crucial in preventing excessive coagulation of soil particles into large clods.

Soil Organic Matter

Dead organic matter in soil not only influences pore size, but itself serves as a sponge that soaks up and retains moisture. Through its decomposition by bacteria, it acts also as a source of carbon dioxide, water, and mineral elements. Certain constituents of dead organic matter, such as waxes, fats, lignins, and some proteinaceous materials, resist decomposition and are converted instead into the dark colloidal substance called *humus*. (A colloidal substance consists of particles larger than molecules but small enough to remain suspended in solution.) The physical and chemical properties of humus affect the character of soil out of proportion to its fraction by weight, in part because of the large surface-to-volume ratio associated with particles of such small size. The fraction by weight of dead organic matter in soil is often in the range of 0.4 to 1.1 per cent.

Living soil organisms (other than the roots of plants) make up an even smaller fraction of the mass of soil (around 0.1 per cent or less). These include: bacteria and algae, which play crucial roles in nutrient cycles; fungi, which perform some of the same functions as bacteria in nutrient cycles, as well as being the principal producers of humus; and larger organisms such as mites, millipedes, insects, and worms, which physically cultivate the soil and help break down organic litter.

Soil Chemistry

The chemistry of a soil is governed largely by the properties of its clay particles and the similarly tiny particles of humus. Clay particles are platelike, with a layered internal structure that leaves negative electric charges arrayed on the surfaces of the plates. Humus particles are also negatively charged on their exteriors. These electrical properties account for the characteristic of soil exchange capacity (the ability to retain and exchange cations such as H^+, Ca^{2+}, Mg^{2+}, K^+, and Na^+).

This function is crucial in governing soil fertility. Without the negative charges on particles of clay and humus, the positively charged nutrient cations released to the soil by the decay of organic matter or added to the soil in fertilizer would quickly be leached away, out of reach of plant roots. Bound to the negative charges on clay and humus, however, those ions are made available to plants gradually when they are replaced at the negatively charged sites by hydrogen ions from the soil. There is actually a replacement heirarchy—hydrogen, calcium, magnesium, potassium, sodium—in which each ion listed tends to displace any that appears to its right in the list, if the two are present in equal amount; hydrogen tends to replace all of the metallic ions. If a particular cation is present in very large quantities, however, it can even replace ions to its left in the hierarchy by sheer force of numbers (that is, by mass action).

The magnitude of exchange capacity depends on the number of negatively charged sites that are available in a soil. The unit of measure is milliequivalents per 100 grams, which means the number of milligrams of hydrogen ion (H^+) that will combine with 100 grams of dry soil. Different types of clay have widely varying exchange capacities (in the range of from 10 to 100 milliequivalents/100 g), whereas humus has the greatest of all (from 150 to 300 milliequivalents/100 g). This chemical role of humus is perhaps even more crucial than the roles humus plays in maintaining soil texture and retaining water. The importance of all three functions provides ample reason for viewing with alarm the depletion of soil humus by certain agricultural practices.

A major source of hydrogen ions in the soil is the production of carbonic acid from the solution of carbon dioxide in water. Another source is nitric and sulphuric acid added by polluted rainfall or formed from nitrogen and sulphur compounds produced by decomposition of organic matter or added in fertilizer. Yet another is organic acids exuded by plant roots or produced by decomposition. Excessive concentrations of hydrogen ions (acidic soil) replace nutrient cations on clay and humus colloids faster than the rate at which the nutrients can be taken up by plants; the result is that the nutrients are leached away. Acidic soils are common in humid climates. A shortage of hydrogen ions (strongly alkaline soil), on the other hand, can leave some nutrient ions too tightly bound to clay and humus to be absorbed by plants. Other nutrients, such as iron and manganese, become trapped in compounds that are extremely insoluble in basic solutions.

Crops differ widely in the pH ranges they tolerate. Most of the grains flourish best in slightly acidic soil, potatoes and berries in quite acidic soil, alfalfa and asparagus in neutral soil. When soil is too acidic for the crops desired, it is common practice to neutralize the soil by adding lime.

Classification of Soils

Differences in soils around the world arise from the mineral compositions of the parent materials and from differing climatio conditions, which together influence the organic and inorganic processes of soil development. Several major types of soil-forming regimes have been identified: the main ones are podzolization, laterization, calcification, gleization, and salinization.

Podzolization is the set of processes associated in its extreme form with cool climates, abundant precipitation, and acid upper soil layers strongly leached of mineral nutrients and the oxides of iron and aluminium. The nutrients, oxides, and humus accumulate in the deeper layers. Fungi are the main soil-forming organisms. Those soils are characteristic of northern forests, although they exist in some circumstances well into temperate and subtropical regions.

Laterization is a set of processes associated with the humid tropics and subtropics. High mean temperatures in these regions permit sustained and rapid bacterial action, which minimizes the accumulation of plant litter and humus. In the absence of the organic acids associated with humus, the soil is neutral, rendering the oxides of iron and aluminium relatively insoluble; those oxides accumulate in the upper soil horizons as hard clays and rocklike material called *laterite* (from the Latin for brick), a mixture of $Fe_2O_3.nH_2O$ and $Al_2O_3.H_2O$.

Calcification occurs in climates where evapotranspiration exceeds precipitation. There is little leaching of the metallic cations, and microbial activity is slow, so the soils tend to be alkaline and rich in humus. Calcium carbonate in solution is carried upward from the water table by capillary action in the season of little surface moisture and is left in the upper soil horizons in solid form when the water evaporates.

Gleization is the set of processes characteristic of poorly drained environments in cool or cold climates. The low temperatures permit heavy accumulation of organic matter, and the excessive wetting produces a sticky clay underneath.

Salinization is the regime characterized by accumulation of highly soluble salts in the soil. The situation arises naturally from poor drainage in regions of low precipitation and high temperatures (deserts), and it can be brought about by faulty agricultural practices—irrigating too parsimoniously in dry climates, or using salt-laden irrigation water. Soils in the salinization regime are weakly to strongly alkaline.

The geographical boundaries separating the spheres of influence of the different soil-forming regimes are often not distinct. An enormous variety of soils results from variations within regimes and from overlaps between regimes, and a quite complicated classification scheme and taxonomy for soils has been devised to help soil scientists cope with this diversity. Fig. 17.3. is a schematic illustration of how a few of the principal *soil types* emerge from the soil-forming regimes.

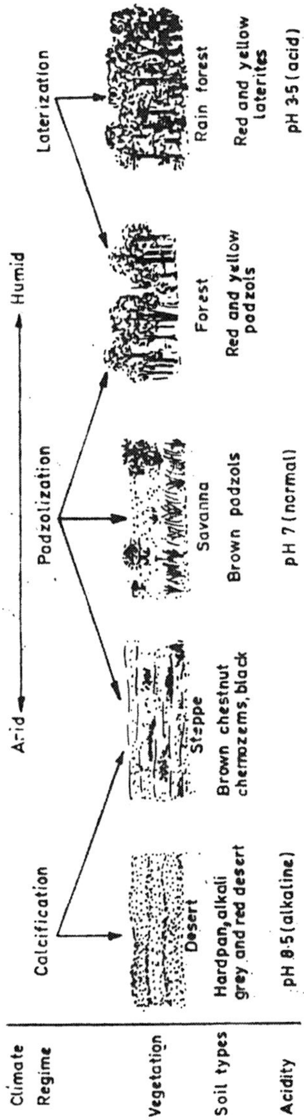

Fig. 17.3. Relation of soil-forming regimes, vegetation, and soil types. (Adapted from Janick et al., 1974)

A somewhat more utilitarian soil classification sch widely used in the United States, is based on the capabilit a soil to sustain various uses—the categories run from Cl (the best soils for agriculture) to Class VIII (suitable only recreation and wildlife).

Soil Microbes

Soils are not just collections of crushed rock; they extraordinarily complex ecosystems. The animals of the are extremely numerous and varied. In forest communitie North Carolina, for example, an estimated 300 million sr invertebrates live in each hactare of soil, about 30,000 square metre. Some 70 per cent of them are mites, a group arthropods that may eventually be found to be as diverse insects. In a study of pasture soils in Denmark, as many 45,000 small oligochaete worms, 10 million nematodes (roun worms), and 48,000 small arthropods (insects and mites) we found in each square metre. Even more abundant are t microflora of the soil. More than a million bacteria one type may be found in a gram of forest soil, as well almost 100,000 yeast cells and about 50,000 bits of fung mycelium. A gram of fertile agricultural soil has yielded ov 2.5 billion bacteria, 400,000 fungi, 50,000 algae, and 30,0 protozoa.

In most natural situations, the plants, animals, an microorganisms of the soil are absolutely essential for i fertility. The beneficial effects of earthworms are well know but most people are unaware of the complex ecological relationships within the soil that make it suitable for the growth o oak trees, sagebrush, corn, or any other plants. The soi contains microorganisms that are responsible for the conversio of nitrogen, phosphorus, and sulphur to forms available to the plants. Many trees have been found to depend on associations with fungi. The fungi get carbohydrates and other essential substances from the roots, and the root-fungus complex is able to extract from the soil minerals that could not be extracted by the root alone. Such mycorrhizal associations are just beginning to be understood, but it is clear that in many areas the visible plant community would be drastically altered if the mycorrhizal fungi were absent from the soil.

Recognizing that most of the complex physical and chemical processes responsible for soil fertility are dependent upon soil organisms, environmental biologists are opposed to the continuing treatment of soils with heavy dosages of deadly and persistent poisons.

Soil Erosion

In the recent past world food output has more than doubled. This was an impressive achievement. But this remarkable feat has a high price. Some of the agricultural practices that boosted food production have also led to excessive soil erosion.

Because of both population growth and rising affluence, world demand for food climbs higher each year. In the face of this continuously expanding demand and the associated increase in pressures on land, soil erosion is accelerating. Soil erosion is a natural process, one that is as old as the earth itself. But today soil erosion has increased to the point where it far exceeds the natural formation of new soil. As the demand for food rises, the world is beginning to mine its soils, converting a renewable resource into a nonrenewable one. Even in an agriculturally sophisticated country like the United States, the loss of soil through erosion exceeds tolerable levels on about 44 per cent of the cropland.

Over most of the earth's surface, the thin mantle of top soil on which agriculture depends is six to eight inches thick. The ever-increasing demand for agricultural products contributes to soil erosion in many ways. Throughout the Third World farmers are pushed onto steeply sloping, erosive land that is rapidly losing its topsoil. Elsewhere, such as the American Midwest, many farmers have abandoned ecological stable, long-term rotations, including hay and grass, as well as row crops, in favour of the continuous row cropping of corn or other crops. In other areas farming has extended into semiarid regions where land is vulnerable to wind erosion when plowed.

The loss of topsoil affects the ability to grow food in two ways. It reduces the inherent productivity of land, both through the loss of nutrients and degradation of the physical structure. It also increases the costs of food production. When farmers lose topsoil they may increase land productivity by the use of chemical fertilizer, or through irrigation to offset the soil's declining water absorptive capacity. Farmers losing topsoil may experience either a loss in land productivity or a rise in costs. But if productivity drops too low or costs rise too high, farmers are forced to abandon their land.

Depletion of soil nutrients and humus need not follow from intensive cropping, as demonstrated by the continued fertility of rice paddies in Asia that have been cultivated continuously for thousands of years. That success has been due to several factors: the high clay content of the soil, the annual augmentation of the alluvial topsoil in the paddies by silt eroded from hilltops, the nitrogen fixation by blue-green algae in the flooded paddies, and the practice of returning human and animal waste to the soils from which the nutrients therein came. Intensive cultivation without either natural or artificial augmentation of nutrients can only exhaust soil fer··· however, and even the use of inorganic fertilizers alone not prevent the depletion of the humus. The practical app tion of the detailed study of soils is to discover both potential and the limitations of different soil types for production of vegetation, and the procedures appropriat using each type as a permanent or renewable resource ra than as a "mine" to be exhausted and abandoned.

The Causes of Soil Erosion

The apparent increase in soil erosion over the past generat is not the result of a decline in the skills of farmers but ratl of the pressures on farmers to produce more. In an integral world food economy, the pressures on land resources are not confined to some particular countries; they permeate the entire world. Many traditional agricultural systems that were ecologically stable recently (around 1950) when there were only 2.5 billion people in the world, are breaking down as world population moves toward 5 billion.

Over the past thousand years as the demand for food increased, farmers devised ingenious techniques for extending agriculture onto land that was otherwise unproductive while still keeping erosion in check and maintaining land productivity. These techniques include terracing, crop rotations, and fallowing. Today, land farmed through these specialized techniques still feeds much of humanity. Although these practices have withstood the test of time, they are breaking down in some situations under the pressure of continuously rising demand.

In mountainous regions such as those in Japan, China, Nepal, Indonesia, construction of terraces historically permitted farmers to cultivate steeply sloping land that would otherwise quickly lose its topsoil. Centuries of laborious effort are embodied in the elaborate systems of terraces in older settled countries. Now the growing competition for cropland in many of these regions is forcing farmers up the slopes. There has been no disciplined construction of terraces of the sort their ancestors built, when population growth was negligible by comparison. Hastily constructed terraces on the upper slopes often begin to give way. These in turn contribute to landslides that sometimes destroy entire villages. For many residents of mountainous areas in the Himalayas fear of these landslides has become an integral part of daily life.

Research in Nigeria has shown how much more serious erosion can be on sloping land that is unprotected by terraces. Cassava planted on land of a 1 per cent slope lost an average of 3 tonnes per hectare each year, comfortably below the rate of soilloss tolerance.

On a 5 per cent slope, however, land planted to cassava eroded at a rate of 87 tonnes per hectare annually—a rate at which a topsoil layer of six inches would disappear entirely within a generation. Cassava planted on a 15 per cent slope led to an annual erosion rate of 221 tonnes per hectare, which would remove all topsoil within a decade. Intercropping cassava and corn reduced soil losses somewhat, but the relationship of soil loss and slope remained the same.

Throughout the Third World increasing population pressure and the accelerating loss of topsoil seem to go hand in hand. In some parts of the world farmers have been able to cultivate rolling land without losing excessive amounts of topsoil by using crop rotations. Typical of these regions is the midwestern United States, where farmers traditionally used long-term rotations of hay, pasture, and corn. Fields planted in row crops, such as corn, are most susceptible to erosion. By alternating row crops with cover crops such as hay, the average annual rate of soil erosion was kept to a tolerable level. Not only do crop rotations provide more soil cover, but the amount of organic matter that binds soil particles together remains much higher than it would under continuous row cropping.

As the demand for foodstuffs increased and as cheap nitrogen fertilizer reduced the need for legumes, American farmers throughout the Midwest, the lower Mississippi Valley, and the Southeast abandoned crop rotations to grow corn or soybeans continuously. The risks associated with this shift in cropping patterns have long been known. Research undertaken in Missouri during the thirties showed an increase in soil erosion from 0.7 tonnes per hectare annually when land was in a corn-wheat-clover rotation to 4.9 tonnes per hectare when the same land was planted continuously to corn. The lower rate is within established erosion tolerance levels, whereas the higher rate would lead to the loss of one inch of topsoil in less than a decade.

Fallowing has permitted farmers to work the land both in semiarid regions and in the tropics, where nutrients are scarce.

In vast semiarid areas—such as Australia, the western Great Plains of North America, the Anatolian plateau of Turkey, and the drylands of the Soviet Union—where there is not enough moisture to support continuous cultivation, alternate-year cropping has evolved. Under this system land is left fallow without a cover crop every other year to accumulate moisture. The crop produced in the next season draws on two years of collected moisture.

In some situations this practice would lead to serious wind erosion if strip-cropping were not practiced simultaneously. Alternate strips planted to crops each year serve as windbreaks for the fallow strips. This combination of fallowing and strip-cropping permitted wheat production to continue in the western U.S.

In the tropics—such as parts of Africa south of the Sahara, Venezuela, the Amazon Basin, and the outer islands of Indonesia—fallowing is used to restore the fertility of the soil. In these areas more nutrients are stored in vegetation than in the soil. When cultivated and stripped of their dense vegetative cover, soils of the humid tropics quickly lose their fertility. In response to these conditions, farmers have evolved a system of shifting cultivation (also known as jhum cultivation in tribal areas of India). They clear and crop land for two, three, or possibly four years and then systematically abandon it as crop yields decline. Natural vegetation soon takes over the abandoned field. Moving on to fresh terrain, farmers repeat the process. When these cultivators return to their starting point after 20-25 years, the soil has regained enough fertility to support crop production for a few years.

Another source of accelerated soil erosion in recent years has been the shift to larger farm equipment, particularly in the Soviet Union and United States. In the United States, for example, the shift to large-scale equipment has often led to the abandonment of field terraces constructed to reduce runoff on sloping lands. In dryland farming regions, tree shelter belts that interfere with the use of large-scale equipment have also been removed. The enlargement of fields to accommodate huge tractors and grain combines also reduces border areas that have traditionally served as checks on erosion.

This transformation of agricultural practices has been fueled by the growing worldwide demand for U.S. feed crops, particularly corn and soyabeans, and by the availability of cheap chemical fertilizer. The rise in demand, in turn, has been amplified by population growth that has hastened the deterioration of traditional agriculture in many countries. As a result, agricultural systems throughout the world are now experiencing unsustainable levels of soil loss.

Assessment of the Problem

The first assessment of world soil erosion was carried out by geologist Sheldon Judson who estimated in 1968 that the amount of river-borne soil sediment carried into the oceans had increased from 9 billion tonnes per year before the introduction of agriculture, grazing, and other activities to 24 billion tonnes per year.

Data compiled in 1980 by three Chinese scientists working for the Yellow River Conservancy Commission in Beijing indicated that river was carrying 1.6 billion tonnes of soil to the ocean each year. (Table 17.1). Hydrologists estimate that an average one-fourth of the soil lost through erosion in a river's watershed actually makes it to the ocean as sediment. The other three-fourths is deposited on foot-slopes, in reservoirs, on river floodplains or other low-lying areas, or in the riverbed itself, which often causes channel shifts.

The Ganges of India deposits 1.5 billion tonnes of soil into the Bay of Bengal each year. The Mississippi, the largest U.S. river, carries 300 million tonnes of soil into the Gulf of Mexico each year, far less than the Yellow or the Ganges. Yet it represents topsoil from the agricultural heartland and is thus a source of major concern for U.S. agronomists.

TABLE 17.1. **Sediment Load of Selected Major Rivers**

River	*Countries*	*Annual Sediment Load (million tonnes)*
Yellow	China	1,600
Ganges	India	1,455
Amazon	several	363
Mississippi	United States	300
Irrawaddy	Burma	299
Kosi	India	172
Mekong	several	170
Nile	several	111

Source: S.A. El-Swaify and E.W. Dangler, "Rainfall Erosion in the Tropics: A State of the Art," in American Society of Agronomy, *Soil Erosion and Conservation in the Tropics* (Madison, Wisc.: 1982).

Scientists have recently documented that vast amounts of wind-borne soil are also being deposited in the oceans as sediment. Island-based air sampling stations in the Atlantic, along with recent satellite photographs, indicate clearly that large quantities of soil dust are being carried out of North Africa over the Atlantic. Visible from satellites, these huge plumes of fine soil particles from the arid and desert expanses of North Africa at times create a dense haze over the eastern Atlantic. Estimates of the amount of African soil being carried west in this way, reported in four studies between 1972 and 1981 (Prospero, 1981), range from 100-400 million tons annually, with the latest report being at the upper end of the range.

Air samples taken at the Mauna Loa Observatory in Hawaii from 1974 through 1982 indicated a continuous movement of soil particles from the Asian mainland, with a peak annual flow consistently occurring in March, April, and May, a time that coincides with a period of strong winds, low rainfall, and plowing in the semiarid regions of North Asia. Scientists at Mauna Loa can now tell when spring plowina starts in North China (Parrington, 1983).

India is one of the few other countries to compile a national estimate of soil loss. In 1975, Indian agricultural scientists collected data on local soil erosion from each of the research stations in the national network maintained by the Indian Council for Agricultural Research. Using these figures, they estimated that 6 billion tonnes of soil are eroded from India's croplands each years (Tejwani, 1982). From this and from an estimate that 60 per cent of the cropland is eroding excessively, the excessive topsoil loss can be calculated by subtracting from the total a tolerance level of five tonnes per acre. This yields an excessive topsoil loss from Indian cropland of 4.7 billion tonnes per year, more than twice the U.S. level. This estimate rests on far less data than does the figure for the United States but it is based on information from agricultural scientists familiar with local soils and it is corroborated by data on siltation of hydroelectric reservoirs, river sediment loads, and other indirect indicators.

As in the United States, soil erosion in the Soviet Union has been spurred by the shift to large, heavy equipment and the enlargement of fields, which eliminated many natural boundary constraints on erosion of soil by both wind and water. Each year an estimated half-million hectares of cropland are abandoned because they are so severely eroded by wind that they are no longer worth farming.

Like the United States, the Soviet Union has an extensive dryland farming area and a substantial irrigated area. The European Soviet Union, which accounts for a large share of total farm output, has moisture levels similar to the U.S. Midwest. In terms of rainfall intensity, topography, and erodibility of prevailing soil types, nothing indicates that soil erosion in the Soviet Union would be markedly less than in the United States Table 17.2. Where cropping patterns are concerned, the Soviet Union relies much more heavily on small grains, whereas the United States relies relatively more on row crops, such as corn and soyabeans.

TABLE 17.2. **Estimated Excessive Erosion of Topsoil From World Cropland**

Country	*Total Cropland (million acres)*	*Excessive Soil Loss (million tonnes)*
United States	421	1,700
Soviet Union	620	2,500
India	346	4,700
China	245	4,300
Total	1,632	13,200
Rest of World	1.506	12,200
Total	3,138	25,400

Source: Worldwatch Institute estimates.

When most of the topsoil is lost on land where the underlying formation consists of rock or where the productivity of the subsoil is too low to make cultivation economical it is abandoned. More commonly, however, land continues to be plowed even though most of the topsoil has been lost and even though the plow layer contains a mixture of topsoil and subsoil, with the latter dominating. Other things being equal, the real cost of food production on such land is far higher than on land where the topsoil layer remains intact.

Indirect Effects of Erosion

When farmers lose topsoil they pay for it in reduced soil

fertility, but unfortunately the costs of erosion are not confined to the farm alone. As soil is carried from the farm by runoff, it may end up in local streams, rivers, canals, or irrigation and hydroelectric reservoirs. The loss of topsoil that reduces land productivity may also reduce irrigation, electrical generation, and the navigability of waterways.

In India, scores of hydroelectric and irrigation reservoirs have been constructed, many of them with assistance from international development agencies. The combination of watershed deforestation and steep slopes being cleared for cultivation is yielding record siltation rates. The siltation rates are now several time as high as the rate that was assumed when the projects were designed (Table 17.3). Not only is the life expentancy of these projects being reduced, but in most cases there will be no alternative sites for dams once the existing ones are rendered useless. A dam site is often unique. Once lost, it cannot be replaced.

TABLE 17.3. **Siltation Rates in Selected Reservoirs in India**

Reservoir	*Assumed Rate*	*(in acre-feet)*	*Observed Rate*
Bhakara	23,000		33,475
Maithon	684		5,980
Mavurakshi	538		2,000
Nizam Sagar	530		8,725
Panchet	1,982		9,533
Ramganga	1,089		4,366
Tungabhadra	9,796		41,058
Ukai	7,448		21,758

Sources: S.A. El-Swaify and E.W. Dangler, "Rainfall Erosion in the Tropics: A State of the Art," in American Society of Agronomy, *Soil Erosion and Conservation in the Tropics* (Madison, Wisc,: 1982), and Center for Science and Environment, *The State of India's Environment 1982* (New Delhi: 1982).

The third major indirect effect of soil erosion is the loss of navigability. Perhaps the most dramatic case occurs in the Panama Canal. The combination of deforestation and the plowing of steeply sloping land in the watershed area by landless people is leading to an unprecedented siltation of the lakes that make up part of the Canal. If the trends of the late seventies and early eighties continue, the capacity of the Panama Canal to handle shipping will be greatly reduced by the end of the century, forcing many ocean-going freighters that have relied on its 10,000 mile shortcut to make the trip via Cape Horn.

Within the United States, soil once used in the Midwest to grow corn now clogs the Mississippi waterways. The Army Corps of Engineers is dredging the inland waterways, particularly in the lower Mississippi River. Vast quantities of soil reach the Gulf of Mexico to become ocean sediment, but substantial amounts are deposited on the way, making large-scale dredging necessary.

Soil Conservation

Eager to maximize food output today, we are borrowing from tomorrow. The loss of over 25 billion tons of topsoil from our cropland each year is the price we pay for shortsighted agricultural policies designed to boost food output at the expense of soils, and of failed or nonexistent population policies.

Soil erosion is a physical process, but its consequences are economic. As soils are depleted through erosion, the productivity of labourers working the eroding land becomes more difficult to raise. In agrarian societies, deterioration of this resource base makes it more difficult to raise income per person. Further, as growth in food output slows, so does overall economic output. In largely rural, low-income societies with rapid population growth, this can translate into declining per capita income, as it already has for a dozen countries in Africa.

Over the long term, world agricultural trade patterns and the international debt structure will be altered. As soils are depleted, countries are forced to import food to satisfy even minimal food needs. Scores of countries in the Third World and Eastern Europe find their international indebteness further aggravated by their chronic dependence on imported food. And the loss of topsoil will force an energy-for-topsoil substitution as it increases the need for fertilizer and fuel for tillage. Other things being equal, land with less topsoil requires more energy to produce our food.

Soil erosion will eventually lead to higher food prices, hunger, and quite possibly, persistent pockets of famine. Although the world economy has weathered a severalfold increase in the price of oil over the past decade, it is not well equipped to cope with even modest rises in the price of food. Although the immediate effects of soil erosion are economic, the ultimate effects are social. When soils are depleted and crops are poorly nourished, people are often under-nourished as well.

Soil Pollution

The effect of pollutants on soil are difficult to evaluate. *Air borne* pollutants emitted by factory stacks travel long distances and slowly deposit on soil. Sulphur oxides present in the fumes are responsible for acid rains and consequent lowering of soil pH. Chlorine and nitrogen oxides are other common gaseous pollutants which combine with water and pollute the soil. Acid rain has been described earlier in the air pollution section. Particulate matter particularly near cement factories, coal tran-shipments, mining belts, etc. reach soil surface of neighbouring regions.

Loffler (1974) has reported that magnesite dust affect very adversely the soil property such as rise of pH, decrease in exchangeable K, Ca, Mg and available P and K to almost a critical level.

A number of chemicals have now become esssential parts of

agriculture. Pesticides are now regarded as the number one pollutant on a global scale. In many countries heavy dose of pesticides are used very often. These, indeed prevent crop and vegetable infection and help in increasing production, but their pollution effect to soil is too high. We have to be very selective and careful in deciding the minimal dose and about the mode of application and to strike a balance between the advantages and disadvantages. DDT reaches the human system via soil and water. In India, the illiterate and uninformed farmer takes a handful of pesticide and throws over the crop unmindful of inhaling the chemical and ingesting it through his contaminated hands. Some of the pesticides are very persistant and non-biodegradable and therefore remain in food chain over long deriods of time. Under normal agricultural conditions, chlorinated hydrocarbons seem to persist for three to five years, whereas organophosphates and carbamates are gone in one to three months or less.

Considerable evidence already exists that the use of insecticides may reduce soil fertility, especially in woodland soils, which are subject to spraying but not artificial cultivation. Populations of earthworms, soil mites, and insects are dramatically changed, and they in turn affect the soil fungi, which are their principal food. Even if bacteria were not affected directly, there is no question that the general effects on the soil ecosystem would carry over to them and to other microorganisms. But it would be unwarranted to assume that the bacteria were not directly affected. It is known that a few microoganisms can degrade DDT to DDD under the proper conditions; a few can degrade dieldrin to aldrin and several other breakdown products of unknown toxicity.

REFERENCES

Janick, J., R.W. Schery, F.W. Woods, and V.W. Ruttan. 1974 *Plant Science*. W.H. Freemam and Company, San Francisco.

Loffler, A. 1974. Researches on the influence of magnesite and fluorine Pollution on the soil, Report, Forest Res. Institute Zvolen (CSSR).

Parrington, Josef R. 1983. "Asian Dust: Seasonal Transport to the Hawaiian Islands," *Science*, April 1983.

Prospero, J.M. 1981. Atmospheric Transport of Soil Dust from Africa to South America, *Nature*, Feb. 1981.

Tejwani, K.G. 1982. Land Use Consultants International, New Delhi, private communication, July 3 1983; Centre for Science and Environment, *The State of India's Environment 1982* New Delhi: 1982).

PART III

POLLUTION CONTROL

18

Controlling Air Pollution

Problem Areas

Few environmental problems have been fraught with so many complexities as those involving air pollution control. While some air pollution abatement efforts have been successful, for example, the reduction of particulate emissions—mainly soot and certain other substances—from coal-and oil-fired furnances, many of the more difficult problems have not yet been controlled. We have permitted machines, factories, and other human activities of various kinds to pour hundreds of foreign substances into one of the most fragile segments of the environment while having only vague ideas about their total effects. Air pollutants tend to be complex mixtures of chemical substances. They often undergo chemical changes in micro-seconds or several months after release into the atmosphere. The toxic hazards of many of them have not been completely studied and are not well understood.

Automobiles, though only one part of the overall problem, illustrate the frustrations that can be encountered in cleaning up auto exhaust. Modifications of the internal combustion engine to reduce one kind of emission led to an increase in another kind of emission almost as obnoxious. The emission control devices such as the catalytic converter, add greatly to the cost of the automobile thereby keeping outworn, high polluting automobiles on the highways.

Air quality standards set by the Environmental Protection Agency (EPA) under the Clean Air Act of 1970 do not intend to *eliminate* air pollutants. Their goal is to bring the concentrations of emissions down to tolerable levels. Deciding tolerable levels, in itself, causes a controversy. Some auto manufacturers claim the standards are too tough, while some authorities contend they are not tough enough.

Electric power plants are among the most troublesome air polluters. Nuclear power plants, low in air pollution emissions, are being built less rapidly than had been anticipated by planners due to public fear of other nuclear hazards from their operation. Thus, fossil fuel emissions from power plant stacks are increasing at a faster rate. Part of the power plant emission problem can be alleviated by burning low-sulphur fuel oil or natural gas and, where these measures are not feasible, by the installation of equipment to remove sulphur oxides and sulphuric acids. This is an expensive procedure, especially where high-sulphur coal is used. In general, nitrogen oxides remains unabated. Another major solution is to build power plants far removed from urban areas.

Assessment of Toxicity

Attempts to regulate the emission of atmospheric toxicants have necessitated the determination of safe atmospheric concentrations. In relation to the industrial environment, these may be established by reference to threshold limiting values (TLVs), formerly known as maximum allowable concentrations, (MACs). The threshold limiting value of a toxicant, or mixture of toxicants, is defined as the maximum concentration to which it is believed healthy workers may be repeatedly exposed without ill effect, on the basis of an eight hour working day. In relation to the atmosphere at large, the threshold limiting value is replaced as a reference point by the 'three minute mean concentration'. This is defined as the maximum permissible concentration of a toxicant, or mixture of toxicants, averaged over three minutes, under the conditions most likely to favour high concentrations at ground level beside a stationary source. The three minute mean concentration is often calculated as a fraction of the threshold limiting value, usually one-thirtieth or one fortieth. Where there is a risk of accumulation of a toxicant, e.g., a heavy metal, an absolute mass emission limit may be established. The absolute mass emission limit is the total amount of toxicant that may be emitted from a stationary source per hour, and which may not legally be exceeded.

The most commonly used assessment of toxicity is the measurement of short term lethality. For a given substance, this involves determining the median concentration which is lethal to a fixed proportion, usually 50 per cent of a test population of organisms after continuous exposure for a fixed time, usually 48 or 96hrs. This is interpolated from the sigmoidal dose response curve (Fig. 18.1) which is obtained by plotting percentage mortality at each dose against dose. The lethal concentration (*LC*) for 50 per cent of a population after continuous exposure for 48h is referred to as the 48h LC_{50} or LD_{50} (*LD*=lethal dose). This abbreviation may be altered appropriately to correspond to different exposure times and proportions of the population. Usually the concentration referred to is that in aqueous solution but it may also be a concentration in air. If the test compound is insoluble or sparingly soluble in water, it must be uniformly dispersed for a repeatable LC_{50} to be obtained. If an emulsifier or other solubilizing agent is used for this purpose, it should be chosen carefully to ensure a minimal contribution to the toxicity of the system.

In medical usage the LC_{50} frequently refers to an injected concentration, and this must be borne in mind when attempting to extrapolate from such studies. Lethal concentrations have been expressed in a variety of units, most frequently in milligrams per litre or per kilogram body weight, but it is desirable that a more uniform approach should be adopted. Where possible, molarity (moles per litre) or molality (moles per kilogram) should be used as this will give a uniform chemical basis for comparing toxicity. The value of the LC_{50} as a

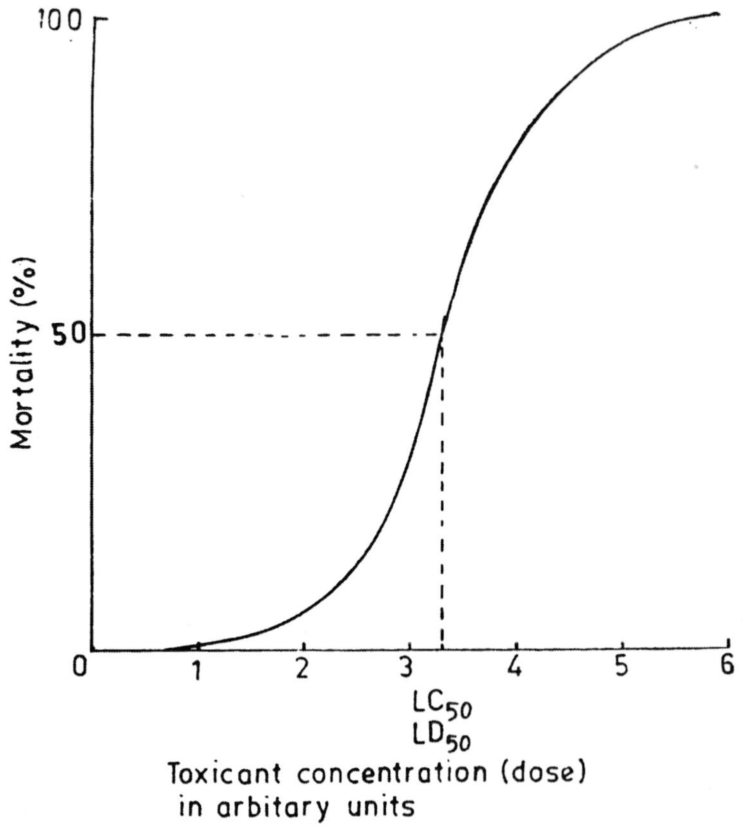

Fig. 18.1 Graph showing the relationship between mortality and toxicant concentration, or dose, following continuous exposure of organisms for a short fixed time, usually 48 h or 96 h.

measure of toxicity has been questioned and it is important to realize that it may be readily misinterpreted. Firstly, it is not constant. Considerable variation is observed, even in a given laboratory with the same population under the same conditions and the same investigator. Secondly, there is no simple correlation between organisms permitting direct extrapolation, even between organisms of the same Class such as rats and man. Thirdly, it gives no indication of the concentrations at which sublethal harm may occur. It is quite possible that a toxicant may be lethal only at very high doses but produce significant ill effects at relatively low concentrations. The most information the LC_{50} gives is an idea of the order of magnitude of the lethal dose under specified conditions. However, it is relatively quick and cheap to determine and provides a basis for initial assessment of the likely hazard from a toxicant and of the effects of various parameters on its toxicity.

Apart from the LC_{50} a number of other figures may be derived from studies of short term lethality. For example, the minimum lethal dose (*MLD*) is the dose which will kill at least one organism in the test population over the test period. This may be described as the LD_1. Another possibility, suggested by Luckey and Venugopal (1977), is that the concept of potential toxicity (p*T*) be introduced on an analogy with pH, i.e., $pT = -\log [T]$ where $[T]$ is the molal concentration of the toxicant and the logarithm is to the base 10. A toxicant with a 24 h LD_{50} of 0.001 mol kg^{-1} would have a 24 h $[T_{50}]$ of 10^{-3} and a 24h pT_{50} of 3. The calculated p*T* corresponds directly to the effect of toxicants on the test population. A small pT_{50} should indicate a relatively harmless substance while a large pT_{50} will indicate a highly toxic substance.

Measurement of short term lethality is one aspect of assessment of acute toxicity. Acute toxicity has been defined as 'the total adverse effects produced by a toxicant when administered as a single dose' (Hunter and Smeets, 1977). However, alternative definitions are used, e.g., 'the total adverse effects produced by a toxicant when given in a single dose or multiple doses over periods of 24 hours or less. Another parameter may be defined, i.e., the EC_{50} or ED_{50} which is the concentration or dose which produces a specific ill effect in 50 per cent of the organisms tested. The EC_{50} is subject to the same provisions given for the LC_{50}.

In July 1970, the Environmental Protection Agency (EPA) of USA was set up and in April 1971, EPA announced the standards for major air pollutants. They are given in Table 18.1. Two levels of achievement were set by EPA, designated as primary and secondary standards. Primary standards are levels of air quality with an adequate margin of safety to protect the public from adverse health effects. Secondary standards, which are stricter, provide protection from other, less serious consequences and include aesthetic factors.

Air Pollution Control Techniques for Smokestacks

Building tall smokestakes was one of the first methods used in an attempt to control industrial smoke pollution. They are quite effective in the immediate vicinity of the industrial plant, provided that there are no temperature inversions. People living downwind from them do not fare so well. The use of cleaner fuels was also an early control method for both home and industrial smoke control. Pollution has been abated at

TABLE 18.1. **National Air Quality Standards, USA**

	Primary	*Secondary*
Suspended particulates		
Yearly geometric mean	75 $\mu g/m^3$	60 $\mu g/m^3$
24-hour maximum	260 $\mu g/m^3$	150 $\mu g/m^3$
Sulphur dioxide		
Annual mean	0.030 ppm	0.022 ppm
24-hour maximum	0.135 ppm	0.097 ppm
Carbon monoxide		
8-hour average	9 ppm	9 ppm
1-hour maximum	35 ppm	35 ppm
Nitrogen oxides		
Annual mean	0.05 ppm	0.05 ppm
Photochemical oxidants		
1-hour maximum	0.08 ppm	0.08 ppm
Hydrocarbons		
3-hour maximum	0.24 ppm	0.24 ppm

times by a change in the manufacturing process or by the replacement of worn-out equipment. It has been suggested that power plants be located near the source of fuel, however, there are difficult problems in transporting electricity over long distances.

Perhaps the most common method of controlling industrial air pollution today is the addition of devices in smokestacks to remove pollutants before they escape into the air. The type used varies with the pollutant. Generally, particles greater than 50 micrometres in diameter may be removed satisfactorily by *cyclone separators* or *wet scrubbers*, whereas particles smaller than 1 micrometre in diameter are handled effectively by *electrostatic precipitators* and *fabric filters*.

Cyclone Separator

In the cyclone separators (Fig. 18.2) the incoming polluted air forms a vortex from which the particles are removed by centrifugal force and fall to the bottom, where they are collected. It is effective for large particles but not for small ones. Efficiencies on a total weight basis vary from 50 to 95 per cent.

Wet Scrubber

In the wet scrubber, the exhaust gases pass through a fine spray of water which very effectively removes particles. It also removes some gases. Efficiencies vary from 75 to 99 per cent. The wastewater, however, often produces a water pollution problem.

Fabric Filters

More effective in removing small solid particles are fabric filters. Exhaust gases are forced through bags of very fine clothlike mesh which are capable of capturing tiny particles. Removal efficiencies may be as high as 99 per cent. These are used widely, especially in cement, carbon black, clay, and pharmaceutical plants.

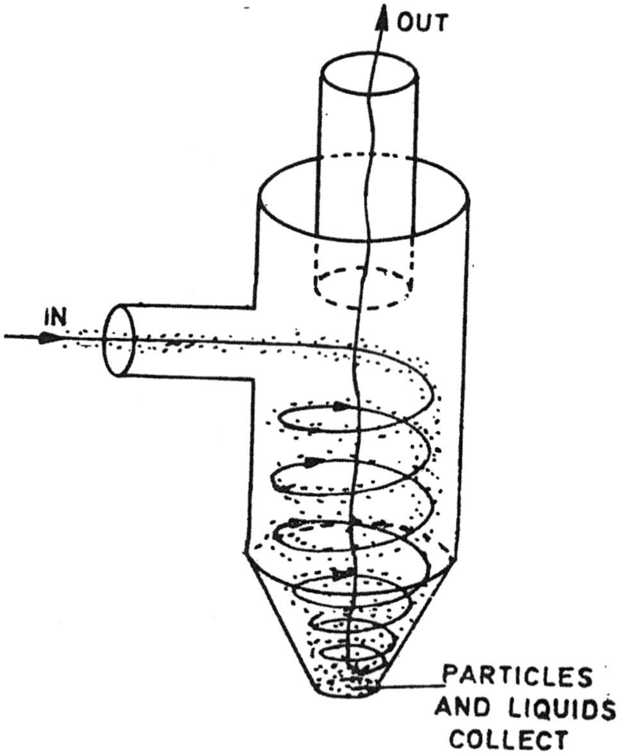

Fig. 18.2. Cyclone Separator to Control Air Pollution in Smoke stacks

Electrostatic Precipitator

The most efficient device for removing particles is the electrostatic precipitator (Fig. 18.3). The incoming waste gases pass over high-voltage electrodes, giving particles an electrical charge. Then they pass between oppositely charged collector plates, to which they are attracted. Vibration of the plates causes the collected matter to fall down into large bins. Collection efficiencies by weight are from 80 to 99.5 per cent. Power plants burning fossil fuels often use electrostatic precipitators and can achieve a removal of more than 99 per cent of all fly ash. The initial cost of the electrostatic precipitator is high, but their power requirements and maintenance costs are small.

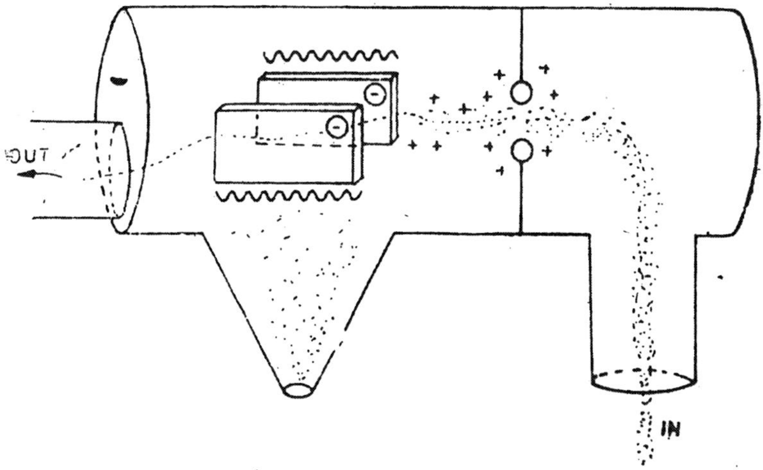

Fig. 18.3. Electrostatic precipitator to control air pollution in smokestacks

Activated Carbon Absorber

The removal of gaseous pollutants is a more difficult problem. They may be absorbed or converted into a less offensive substance. The wet scrubber used to remove particles is also effective in absorbing gaseous pollutants, such as ammonia, hydrogen sulphide, hydrogen chloride, chlorine, and hydrogen fluoride.

Activated carbon is used to remove polluting gases, since it attracts thin layers of gases and vapours to its surface, leaving the carbon unchanged. The carbon, after being saturated with the collected pollutant, can be regenerated quickly with a steam bath. It is used in removing hydrocarbons.

Controlling Gaseous Pollutants

Carbon Monoxide

The main source of anthropogenic carbon monoxide is the

internal combustion engine and so it has become the prime target for control procedures. Since other toxicants are simultaneously released, there is always the possibility that a method of lowering the concentration of carbon monoxide may increase the concentration of these. For example, supplying a stoichiometric air/fuel mixture to the internal combustion engine will give low carbon monoxide and hydrocarbon emissions, but high emissions of nitrogen oxides. The principal technique evolved to reduce carbon monoxide emission has been the development of exhaust system reactors, which convert carbon monoxide to carbon dioxide, and hydrocarbons to carbon dioxide and water. There are two types of reactor: those that are catalytic and depend on platinum or palladium, or a mixture of both, and those that use a high temperature chamber. None of the catalysts available is entirely satisfactory. They are all poisoned by lead and sulphur compounds which may be present in gasoline, and by ethylene dibromide which is added to leaded gasolines to prevent build-up of lead deposits in engines. They are also poisoned by organothiophosphate compounds released from lubricating oil. It is even possible that platinum and palladium released from catalytic reactors into the environment may prove to be toxic. Certainly, water soluble platinum salts have harmful effects. High temperature chambers do not have any of these shortcomings but they are more expensive to operate because they increase fuel consumption. Other approaches to reducing carbon monoxide emission include improvements in engine design, development of substitute fuels and development of new power sources. None of these has so far proved entirely satisfactory.

Hydrocarbons

Control of hydrocarbon emission from transportation involves the same treatment of the exhaust fumes as does removal of carbon monoxide. In addition, loss of hydrocarbons by evaporation from the fuel tank and carburettor is reduced by installation of a collection system which eventually returns them to the fuel induction system. Hydrocarbon emissions from stationary sources are controlled by use of afterburning, adsorption, absorption and condensation. Afterburning completes the oxidation of the hydrocarbons to carbon dioxide and water. For adsorption, a bed of activated carbon is used. This bed is periodically cleaned with steam, from which the hydrocarbons may be recovered following condensation. Where possible, the recovered hydrocarbons may be utilized for other processes. Otherwise, they may be disposed of by oxidation to carbon dioxide and water. Absorption is achieved by passing exhaust fumes through a liquid in which the hydrocarbons will dissolve or become suspended. Again, they may be recovered for subsequent use or they may be eliminated by oxidation.

Oxides of Sulphur

Modifying the catalytic units in car exhaust systems to reduce sulphuric acid production may involve changing the method of catalyst use, adding a sulphate trap, or completely removing sulphur from gasoline prior to its use. As far as coal burning, the main source of anthropogenic sulphur oxides, is concerned, there are three possible ways of reducing emission. Firstly, one can use coal with a very low sulphur content. Unfortunately, there are only a few regions where this is economically feasible. Secondly, sulphur can be removed from coal by a water washing process. However, this is expensive and removes only pyritic sulphur. Thirdly, coal can be gasified and the sulphur oxides produced trapped by a suitable chemical reaction. This method looks the most promising, but it is still fairly costly in operation.

Oxides of Nitrogen

Most of the methods employed to reduce anthropogenic emission of nitrogen oxides involve modification of the combustion processes involved. Reduced peak temperature, and/or reduced oxygen availability at this temperature, lowers nitrogen oxide levels and this is the main approach used. Catalytic modification of exhaust fumes is difficult because it must be reductive, whereas removal of carbon monoxide and hydrocarbons is oxidative.

Legislative Control of Air Pollution

England

Political control of air pollution is not a new concept. In 1273, King Edward I of England promoted the passage of a law outlawing the use of certain types of coal. The law, however, was never enforced. In the late fourteenth century, King Richard III put a high tax on coal. Charles II, in 1661, commissioned John Evelyn to make a report on London's air pollution. His ***Fumifugium*** described with great accuracy the effects of London's smoke on respiratory diseases, the reduction of sunshine, and the corrosion of materials. Evelyn's proposals were ignored. London's smoke problem continued, and during the nineteenth century, it became notorious, The Public Health Act of 1875 did contain a smoke abatement section, but because of pressure from industrialists, it was never enforced. At the turn of the century, a number of smoke abatement societies were formed. One was led by H.A. Des Voeux, who coined the word ***smog*** for the fog combined with smoke that occurred so often in London. This type of smog became known as ***classical smog*** in contrast to that caused by automobiles, which is referred to as ***photochemical smog***. Despite some legislation, Britain's smog worsened. It was not until after London's disasterous smog episode of December 1952 that effective legislation was passed. The Clean Air Act of 1956 gave officials the authority to curtail industrial emissions and to restrict domestic heating to smokeless fuels. Coal smoke was dramatically reduced, and London today is a far cleaner city.

United States

Probably the first smoke ordinance in the United States was passed by the City Council of Chicago in 1874. Many of these early laws specified violation in terms of greyness of smoke. California was a pioneer in state air pollution control. In 1947, it passed the Air Pollution Control District Act, providing for countrywide control districts. In 1966, the City Council of New York passed a law requiring more extensive use of low-sulphur fuels and the upgrading of incinerators. They also passed a law banning all smoking in public places after November 1, 1974. Such places include supermarkets, elevators, classrooms, hospitals, and recreational, religious, and social gathering places involving 50 or more persons.

No federal air pollution legislation was passed in USA until 1955, and this only provided for research and for technical assistance to state and local governments. The first important national legislation was the Clean Air Act of 1963, which granted permanent authority to federal air pollution control activities. Amendments followed in 1965, 1967, and 1969, with the most important additions made in 1970. The latter provided for the establishment of national air quality standards and their achievement by July 1975 through implementation plans drawn up by air quality control regions and states. The Environmental Protection Agency (EPA) was set up in July 1970.

The Indian Scene

The Air (Prevention and Control of Pollution) Act, 1981 was promulgated on 16th May, 1981. No separate agency was set-up to monitor and control air pollution, and the Central Board for prevention and control of water pollution also looked after the control of air pollution.

The water (Prevention and Control of Pollution) Act, 1974 became effective from 23rd March, 1974. This Act was ammended in 1988. While amending the Water (Prevention and Control of Pollution) Act, 1974 in 1988, the Central/State Boards for the Prevention and Control of Water Pollution were renamed as Central/State Pollution Control Boards as these Boards also deal with air pollution. Consequent to this amendment, there is now a Central Pollution Control Board (CPCB) that advises the Pollution Control Division of the Ministry of

Environment and Forests, New Delhi. There are also State Pollution Control Boards (SPCB) to advise the respective State governments.

The following legislations now exist for pollution control in the country:

1. The Water (Prevention and Control of Pollution) Act, 1974.
2. The Air (Prevention and Control of Pollution) Act, 1981.
3. The Environment (Protection) Act, 1986, and
4. The Motor Vehicles Act, 1988.

According to the provisions of the Motor Vehicles Act, 1988, from 1-10-1989 every motor vehicle shall comply with the following emission standards:

(i) idling emission limit for four-wheeler petrol driven vehicle shall not exceed 3 per cent by volume.

(ii) idling emission limit for two-wheelers and three-wheelers petrol driven vehicles shall not exceed 4.5 per cent by volume.

(iii) smoke density for all diesel driven vehicles shall be as follows:

Method of test	*Maximum smoke density*		
	Light mL	*Bosch units*	*Hartidge*
Absorption Coefficient			
(i) Full load at a speed of 60% to 70% of maximum engine rated speed declared by the manufacturer	3.1	5.2	75
(ii) Free acceleration	2.3		65

Global Efforts

The real awareness about environmental protection at global level was recognised at the United Nations Conference on the Human Environment held at Stockholm (Sweden) in June 1972. The focus of attention was to tackle environmental problems efficiently. In this conference, the late Prime Minister Indira Gandhi took keen interest and initiative to take appropriate steps for the protection and improvement of the environment.

REFERENCES

Anderson, J.E. 1974. *Environment, 16*(3), 6-11.

Crossland, J. 1974. Cars, Fuel and Pollution, *Environment, 16*(2), 15-27.

Hunter, W.J. and J.G.P.M. Smeets. 1977. *The Evaluation of Toxicological Data for the Protection of Public Health*, Pergamon Press, Oxford.

Luckey, T.D. and B. Venugopal. 1977. *Metal Toxicity in Mammals*, Vol I, Plenum Press, New York.

National Academy of Sciences (NAS), 1975. *Air Quality and Stationary-Source Emission Control*, Washington, D.C.

Stern, A.C. (Ed.). 1968. *Air Pollution*, Vol. I, Academic Press, New York.

Turk, J. and A. Turk. 1984. *Environmental Science*, Saunders College Pub. Philadelphia.

19

Treatment of Polluted Water

Waste water Treatment

Sources and Types of Toxic wastes

Domestic wastes contain ammoniacal nitrogen at levels up to 50-60 mg L^{-1} and may contain sulphide at levels up to 50 mg L^{-1}, both of which can cause damage to aquatic life, if discharged. Agricultural wastes may also contain even higher levels of both ammonia and sulphide plus a range of toxic pesticides and herbicides and the drainage from silage may be toxic due to low pH values. The majority of toxic wastes are associated with industry. Here the range of possibilities is too wide. Table 19.1 gives some indication of the main types of toxic industrial waste and the toxins that they contain.

Pre-Treatment of Toxic Wastes

Pre-treatment of industrial wastes prior to discharge to a sewer is widely practised. In some cases this is to comply with the restrictions imposed by Water Authorities and Local Authorities on wastes entering a sewer. There are also a large number of specific toxic substances whose discharge to sewers is controlled.

TABLE 19.1. **Sources of some Common Toxins**

Toxin	*Sources*
Acids—mainly inorganic but some organic causing pH < 6	Acid Manufacture, Battery Manufacture, Chemical Industry, Steel Industry
Alkalis—causing pH > 9	Brewery wastes, Food Industries Chemical Industry, Textile Manufacture
Antibiotics	Pharmaceutical Industry
Ammoniacal nitrogen	Coke Manufacture, Fertiliser Manufacture, Rubber Industry
Chromium—mainly hexavalent but also less toxic trivalent form	Metal processing, Tanneries
Cyanide	Coke production, Metal Plating
Detergents—mainly anionic but some cationic	Detergent Manufacture, Textile Manufacture, Laundries, Food Industry
Herbicides and Pesticides—mostly chlorinated hydrocarbons	Chemical Industry
Metals—mainly copper, cadmium cobalt, lead, nickel, mercury and zinc	Metal processing and plating, Chemical Industry
Phenols	Coke Production, Oil Refining Wood Preserving
Solvents—mostly benzene, acetone, carbon tetrachloride, alcohols	Chemical Industry, Pharmaceuticals

The main advantages of treatment on site are the possibilities of recovering heat or specific substances in an uncontaminated condition and the economies which might result from treatment at higher temperatures and/or higher concentrations. Where the toxic materials are organic in nature there is often some difficulty in treatment due to inhibition of bacteria growth. It is often easier and cheaper to develop the necessary bacterial flora in an *on-site* treatment plant.

This to some extent depends upon the concentration and toxicity of the substances concerned. In some cases dilution of the waste by admixture with sewage reduces the toxic inhibition making it preferable to treat the industrial waste and sewage together. Also many industrial wastes are deficient in some nutrient such as nitrogen or phosphorus. The desirable ratio of BOD: N:P is 100: 5 : 1 and the ratio in domestic wastes is commonly 100 : 18 : 2.5 so that deficiencies in industrial wastes can be balanced.

Even where it is decided not to carry out pre-treatment of the toxic waste by chemical or biological methods it is often useful to install devices to improve the effluent quality by simple physical means. This includes some form of screening to reduce solids. Also some form of balancing to reduce variations in concentration, flow, pH etc. and some traps to prevent the escape of oil and grease and some grit arrestors.

The processes which are used may be classified as physical, chemical and biological. The physical processes are summarised in Table 19.2. Where the toxic wastes contain or are composed of organic materials it may also be necessary to provide some biological treatment especially if the effluent is to be discharged directly into a water way. Many different types of processes are used but the following are the most popular.

(a) High-rate filtration using plastic media with very high rates of recirculation.

(b) Activated sludge using contact stabilization.

Like all biological processes these can suffer from toxicity problems especially where the concentration of toxin is not constant. In general terms it is easier for bacteria and other microorganisms to adapt to toxic substances than for organisms like worms, fly larvae etc. For this reason conventional percolating filters have not proved successful—the lack of grazing fauna has led to persistent ponding.

TABLE 19.2. **Physical Methods of Pre-Treatment**

Process	*Aim*	*Examples*
Screening	Removal of coarse solids	Vegetable canneries, Paper Mills
Centrifuging	Concentration of solids	Sludge dewatering in chemical industry
Filtration	Concentration of fine solids	Final polishing and sludge dewatering in chemical and metal processing
Sedimentation	Removal of settleable solids	Separation of inorganic solids in ore extraction, coal and clay production
Flotation	Removal of low specific gravity solids and liquids	Separation of oil, grease and solids in chemical and food industry
Freezing	Concentration of liquids and sludges	Recovery of non-ferrous metals
Solvent Extration	Recovery of valuable materials	Coal Carbonizing and Plastics manufacture
Ion Exchange	Separation & Concentration	Metal processing
Reverse Osmosis	Separation of dissolved solids	Desalination of process and wash water
Adsorption	Concentration and removal of trace impurities	Pesticide manufacture, dyestuffs removal

Due to a combination of high organic strength and inhibition from toxic substances it is unusual to obtain complete treatment of toxic industrial wastes by conventional primary and secondary treatment. The effluent from high-rate filters often has a BOD and COD similar to settled sewage and is either suitable for discharge to a sewer or for further biological treatment on site.

Chemical treatment of industrial wastes may be used in addition to and to some extent in place of biological treatment. The aims are somewhat different since biological treatment is mainly a way of oxidising organic matter or a way of converting it into a settleable form. Chemical treatment can also provide oxidation through chlorine, ozone, etc., but this is used only for oxidising particular compounds like cyanide since it is expensive and liable to lead to the production of undesirable chlorinated organics. It is mainly used for pH correction and improving the removal of solids. The commonest chemicals in use are shown in Table 19.3.

TABLE 19.3. **Chemicals used in Industrial Waste Treatment**

Chemical	*Purpose*
Calcium hydroxide	pH adjustment, precipitation of metals and assisting sedimentation
Sodium hydroxide	Used mainly for pH adjustment in place of lime
Sodium carbonate	pH adjustment and precipitation of metals with soluble hydroxide
Carbon dioxide	pH adjustment
Aluminium sulphate	Solids separation
Ferrous sulphate	Solids separation
Chlorine	Oxidation

Primary and Secondary Treatment

If the pre-treatment of toxic industrial wastes is successful, no difficulties should be encountered in subsequent treatment. However, no pre-treatment system is perfect and malfunction will occasionally occur mostly due to variations in the manufacturing process. As a result toxic material together with possible overload of organics and solids may be passed on to the subsequent treatment stages.

The effect of toxic materials on primary sedimentation is insignificant, since this is a purely physical process of sedimentation and flocculation. However the effect of the primary sedimentation on toxic wastes can be very important. Toxic materials in suspension such as particulate metals are effectively removed. Also the flocculant material has a great capacity for adsorption and removes the majority of dissolved metals, pesticides and other toxic organics. In one respect this is beneficial since it renders the waste material less inhibitory for biological treatment but it selectively concentrates the toxins in the sludge and may give rise to problems in digestion and in sludge disposal.

The key to successful secondary treatment of wastes containing toxic materials is the adaptation of the microorganisms to the presence of the toxins. Bacteria and to a lesser extent protozoa show considerable ability to acclimatise to the presence of toxic substances and a great adaptability in degrading new synthetic organics. Metazoa are rather less adaptable and for this reason activated sludge is generally better than percolating filters for treating toxic wastes. Experience with treating toxic industrial wastes on percolating filters has shown frequent ponding problems due to a lack of activity by the grazing fauna. High-rate filters which utilise hydraulic scouring for film control have been used successfully and the high recirculation ratios help to dilute any incoming toxins. This helps to overcome the other disadvantage of filters which is due to the plug-flow nature of the process. Any shock loads of toxin are not as readily diluted as in a completely mixed reactor.

The activated sludge process is generally preferred for dealing with wastes containing an admixture of toxic materials. In particular the completely mixed version gives immediate dilution of any shock loads. The dangers with activated sludge are that:

(a) The toxin may reach concentration which inhibits enzyme activity.

(b) Some toxins also affect the bacterial surface and therefore affect settleability.

Nitrification is particularly sensitive to these problems. It is therefore important in treating a toxic waste by either biological process that a microbial population is developed which is acclimatised to the presence of the toxin and in the case of degradable toxins contains sufficient numbers of organisms which can metabolise the toxins. These twin aspects of acclimatisation require great care in the start-up operation and may need a period of several months before successful operation is achieved.

Sludge Treatment and Disposal

Toxins which remain in solution during primary and secondary sedimentation do not appear in the sludges and thus cause no further difficulties. Toxicity in digestion may also occur due to soluble toxins in the treatment of industrial wastes by anaerobic methods. But the main problems are due to toxins which are in settleable form or are readily adsorbed, are selectively concentrated in the sludge and give rise to difficulties in digestion (and in subsequent disposal). The classes of toxins involved are mainly metals, chlorinated hydrocarbons, organic solvents and detergents.

Sewage Treatment

The processes of treating municipal sewage are broadly classified as *primary*, *secondary*, and *tertiary*. The particular process used in a given situation depends on the volume to be treated, the location of the outfall, the dilution factor, the potential hazard to users receiving the water, and in many cases, the cost of the project.

Primary Treatment

The sewage is often disposed of as raw sewage of effluent

from primary treatment. Primary treatment consists of removing floating and suspended solids by mechanical means (Fig. 19.1.). More than one-half of the suspended solids can be removed by primary treatment. First the large solids are screened out and grease and scum are removed. This is followed by sedimentation in a basin to remove the remaining solids, called *primary sludge*. The screens, called trash racks, consist of steel bars about 2 to 4 inches apart. In some cases sand and other coarse material is removed by *grit chambers* to further protect pumps and other equipment from damage. Sometimes the waste water is then run through fine screens. Usually after screening and removal of grit, the waste water is run directly into settling tanks. The settling tanks may have skimming devices, or the removal of scum may be done separately.

The primary sludge is a burdensome problem because it is bulky and must be removed. It contains 94 to 99 percent water. Usually the first step is to remove as much of the water as possible. In some cases the sludge is dried in beds with some of the water being removed by filtration. The residue is disposed of on land. Because the sludge itself can comprise a pollution problem, a better method is to bring about microbial decomposition in *sludge digestion tanks* before drying. The microbial action occurs under anaerobic conditions, and because this proceeds slowly at low temperatures, the digestion tanks are usually maintained to 25 to 32°C, at which temperatures the sludge decomposes in 20 to 30 days. The digested sludge will have been reduced to about one-third of its original volume and will be relatively inoffensive.

Primary treatment consists of removing floating and suspended particles by mechanical means (Fig. 19.1a). The other products of primary treatment are gases and the fluid or clarified waste water. The gas is mostly methane, which is usually burned as fuel to provide heat for the digesters and other equipment. The clarified waste water has highly objectionable properties and in most cases is put through a secondary treatment.

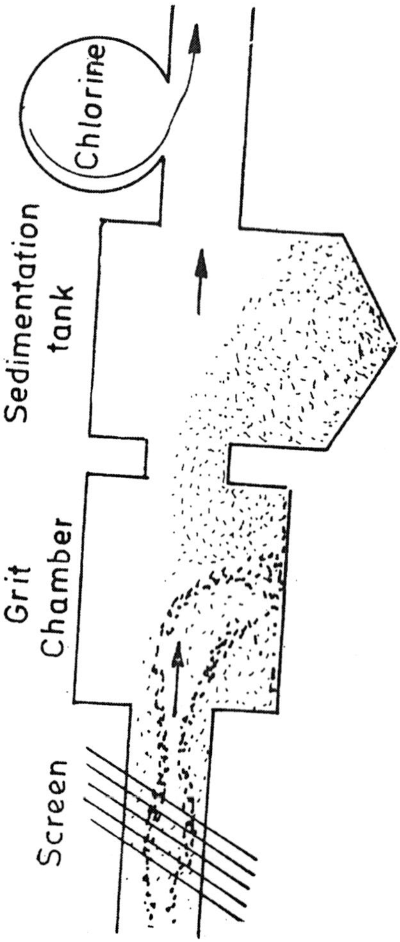

Fig. 19.1(a). Primary Sewage Treatment

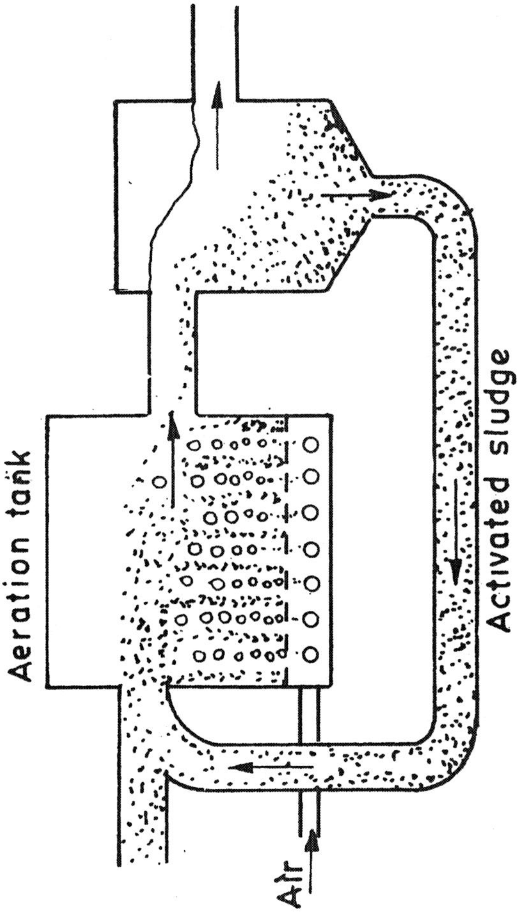

Fig. 19.1(b). Secondary Sewage Treatment

Secondary Treatment

Secondary treatment of waste involves the biological degradation of organic materials by microorganisms under controlled conditions (Fig. 19.1b). The usual method is to bring about the biological oxidation of the organic material under aerobic conditions, in which the waste is aerated to supply oxygen for the microorganisms. The degraded material settles out in secondary settling tanks and is therefore described as being removed by sedimentation. The sediment containing the microbial growths and their by-products is called *secondary sludge* or *activated sludge*. The clarified waste water is discharged in the outfalls to rivers, lakes, bays, lagoons, or oceans. Some of the sludge is returned to aeration tanks where it is recycled with the incoming waste. Most of the sludge, however, must be removed, and this operation accounts for one-fourth to one-half of the operation.

Sludge

Both primary and secondary sludges are usually transferred to sludge digestion tanks where decomposition takes place under anaerobic conditions. Various methods for water removal are used. Because the removal of the sludge is a major part of sewage disposal, a great deal of effort has gone into drying and finding uses for the product, or locating disposal sites and transporting the sludge there.

A few of the treatment plants heat-dry the activated sludge, usually after some form of mechanical water removal, called *dewatering*, and sell the product as fertilizer. The process is costly. It is less expensive to incinerate the sludge in furnaces or to use it as landfill. Incineration sterilizes the sludge and reduces its volume. However, incineration has some disadvantages. It creates an air pollution problem and leaves an ash that must be disposed of. Still, disposing of a small amount of ash is easier than getting rid of a large amount of sludge.

Final Disposal

The methods used for disposal of sludges are usually those that are the least costly. The choice of disposal site whether on land or sea depends on the proximity of the treatment plant to suitable disposal locations. Dewatered sludges are commonly used as landfill. Liquid sludges are disposed of either on land or in bodies of water. Liquid sludge is used to fertilize or condition agricultural land. However, problems of odour, water pollution, and stimulation of insect and algal growth, as well as other aspects of public health and aesthetic values, must be considered.

Tertiary Treatment

Any treatment to further purify waste water beyond the methods commonly employed for secondary treatment may be called "advanced treatment." The term *tertiary treatment* is often used to refer to certain advanced treatment procedures for treating the effluent coming from secondary treatment. The purpose is the removal of the contaminants of waste water that remain after secondary treatment. These contaminants consist mainly of suspended solids, dissolved organic compounds, and the inorganic plant nutrients nitrogen and phosphorus. A thorough job of cleaning up sewage waste water is difficult and costly.

A number of alternative techniques depending upon the nature of pollutant are available. The methods involve removal of nutrients by autotrophic plants, flocculation of colloidal particulate matter, use of adsorbants, use of oxidising agents like $KMnO_4$ or O_3, use of *reverse osmosis* by artificial pressure device across semipermeable membrane, or by specific chemical techniques. Chlorination is done to eliminate orthophosphates remaining in water after primary and secondary treatments which can be precipitated by Ca or some other metal ions. Aluminium sulphate, ferric chloride, ferrous (ic) sulphate and calcium hydroxide are good coagulants. Nitrates are removable with the help of *Nitrococcus denitrificans* and *Thiobacillus denitrificans* which denitrify and release nitrogen and water.

REFERENCES

Anon, 1970. Drinking Water—Is it Drinkable *Env. Sci. and Tech.* 4 (10). 811—813.

Eckenfelder, W.W. and D.J. O'Connor. 1961. *Biological Waste Treatment*. Pergamon Press. New York.

Sawyer, C.N. and P.L. McCarthy. 1978. *Chemistry for Enviromental Engineering*. Mc Graw Hill, New York.

Tearle, K. (Ed.) 1973. *Industrial Pollution Control*. Business Books Ltd. London.

Varshney, C.K. (Ed.) 1985. *Water Pollution and Management*. Wiley Eastern Limited, New Delhi.

20

Solid Wastes and Recycling

Solid Wastes

Litter along the roadsides, trash on the beaches, heaps of refuse in the streets, the pungent piles of rotting garbage are unpleasant reminders of untidy human habits. Thousands of backyards, vacant plots contain the remains of discarded machines, parts of vehicles, food wastes and so on. Open dumps scar the landscape. In some countries solid wastes are dumped into the ocean. The growing mass of solid waste produced annually the world over includes millions of tons of paper and paper products, plastics, billions of bottles, cans, tyres, junked machines and discarded automobiles and millions of other appliances of different sizes and kinds. Probably the United States produces the largest solid waste. The city of New York produces the largest waste around one tonne each household per year. Around 15 per cent of solid waste from every household in an affluent society consists of food wastes. There is considerable loss of raw material and energy which goes into the manufacturing of the material thrown away as solid waste.

Disposal of Solid Wastes

The most commonly used methods for final disposal of solid wastes are sanitary landfill and open dumping. These methods are cheaper than the other disposal methods. The other advantage is that we can avoid the acute pollution problems associated with discharging wastes into waterways or polluting the air from incineration.

If properly planned, landfills can be used later for construction sites or recreational facilities. However, there are major disadvantages to open dumps, which are a public health problem, breeding flies, rats etc. Burning combustible wastes produces mixtures of particulate and gaseous matter that are obnoxious. Land disposal of solid wastes requires large parts of land. The increasing demand of land for human habitations and the ever-growing mountain of solid waste make the problem complex.

The adverse economics of this type of the disposal of solid wastes are far-reaching. Mineral resources such as copper, zinc, lead and tin are limited, and the ores are non-renewable. These metals are not destroyed by disposal but are often scattered beyond recovery. The squandering of these resources that occur when wastes are dumped indiscriminately, will produce a disruptive economic impact at some time in the future unless improvements are adopted. *Recycling* will alleviate some problems associated with the increasing density of population.

Incineration

In most developed countries the residents collect the wastes in polythene bags, and put them at prescribed places at a certain time of the day. Then onwards it becomes the responsibility of the municipal or other local bodies to collect and carry them to appropriate disposal sites. In well organised waste disposal systems articles such as cans, bottles etc. are separated and then the remaining garbage subjected to compaction in special trucks. These are transported to selected garbage disposal sites, like landfills. composting pits, open dumps or to incinerators. Once the garbage is collected, useful or recyclable articles are separated. In India, some people earn their livelihood by selling renewable resources handpicked from garbage. The solid wastes to be incinerated are *highly combustible* trash, like paper, cardboard, plastics or rubber scrap, *combustible* wastes like wood scraps, cartons, floor sweepings, food *wastes* and other combustible articles. The incinerators produce toxic ash which must be landfilled.

The incinerators are so designed that the heat generated during incineration may be usefully utilised. The most common use is to generate steam. This partly recovers the cost of waste collection and disposal. Some waste materials produce corrosive fumes. Hence a careful segregation of such materials is desirable before compaction and incineration.

Recycling

Recycling Through Composting

One of the best places to start recycling is in the home garden with a compost pile. Compost, by definition, represents a mixture of plant material, soil, eggshells, straw, manure, and other materials that can be broken down by microorganisms (bacteria and fungi). Its main value for use in the home garden, after decomposition has converted it into rich humus, is to improve soil texture, increase soil water-holding capacity, serve as a mulch, prevent or retard weed growth, and protect plants from excess winter injury due to low temperatures or desiccation. It may also serve as source of nutrients, especially if ashes and manure have been added to the compost pile. However, the fertilizer value of most compost humus is low; 0.5 per cent nitrogen, 0.4 per cent phosphorus, and 0.2 per cent potassium.

To make good humus we can add the following to a compost pile: eggshells, citrus rinds, coffee grounds, kitchen vegetable garbage, tops of cut perennials, leaves, grass clippings, weedy annuals, ashes, straw, manure, crushed granitic rock, soil, inorganic fertilizer, seaweed, overripe fruit, water, and earthworms.

Such a mixture of garden and home by-products will contain many types of fungi and bacteria which act to decompose the compost substrates. However, to make sure that decomposition occurs satisfactorily, many compost makers add ***compost starter***, which is an aqueous suspension of bacteria and/or fungi that are adept at breaking down vegetable matter. The decomposition action by these organisms works best under aerobic conditions in warmer weather where there is a favourable supply of moisture. Like secondary treatment of sewage, the microbial action breaks down the vegetable and other matter to humus plus CO_2 and, in the process, liberates much heat, generating temperatures typically between 50 and 80°C.

A compost pile may be constructed in many ways but basically, one makes alternating layers of vegetable matter and soil. To stimulate microbial action, some fertilizer and water should be added periodically to the compost pile, especially during periods when the air temperature is warm. The soil is primarily added to provide a source of microorganisms and to hasten the microbial decomposition action by these microorganisms. It also retains moisture needed by these organisms to grow.

Energy Production Through Recycling

Another approach to recycling involves the use of animal wastes (manure) and plant debris (including algae) to produce methane (CH_4) and organic fertilizer for urban, farm, and garden use. Methane (biogas) generators are in widespread use in Germany, France, India, Japan and the United States. They can be used in both northern temperate and in tropical regions. Several of the big advantages to their use are that they utilize readily available substrate; they help to solve energy needs especially when used in combination with wind generators and solar panels; they provide useful fertilizer by-products; and they help to recycle immense quantities of manure and garbage.

In contrast to composting, an aerobic operation, methane generators involve breakdown of organic matter in the absence of oxygen (anaerobic conditions). The latter, therefore, involves a well-known process called fermentation. The main products of this fermentation are CH_4 and CO_2, with trace amounts of hydrogen sulphide and nitrogen gas also given off. The conversion occurs at or around normal atmospheric pressure and at temperatures ranging between 15 and 50°C.

In general, conversion of 450 g of dry organic matter yields between 4.5 to 6.5 cubic feet of methane. In India methane gas generators are called ***biogas plants***.

Recycling Paper and Plastics

The United States recycles about one-fifth of its paper. In contrast, Japan recycles about one-half of its paper. Harvesting trees to make pulp for paper manufacture costs less than the recycling process. Removal of ink is now feasible but expensive, and recycled paper is generally not useful in making white bond or slick paper for magazines. Recycling will become more profitable as pulp becomes more scarce, labour costs increase, and recycled paper comes into greater use.

One of the most emphatic arguments for recycling paper is that it results in a reduction of solid waste and a reduction of the demands placed on a nation's forests and woodlots. The rate at which timber is being cut in forests for lumber and pulp is indeed alarming.

Non-degradable plastics are put in disposal sites. They make a relatively stable material for landfil because they do not breakdown and pollute underground water systems. Some plastics are however degradable. In developing and under-developed countries plastics are generally recycled. However, the quality of recycled plastics is not as good.

Recycling Glass, Metal and Rubber

Various methods have been devised to separate non-ferrous from ferrous metals; glass from metal and different colours of glass. Of course the more economical method of achieving this separation is in individual homes, restaurants and other places where the materials are used. Broken glass is used in the manufacture of more glass bottles. This saves energy—melted glass uses only 2/3rd of heat energy required to produce fresh glass from silica and lime. Metal is easily recycled. Used tyre casing are suitable for reuse in the manufacture of synthetic rubber. Shreaded tyres mixed with rubber-base stock is used for playground surfacing.

Developing Countries

The hi-tech methods used for solid waste management and sewage treatment are highly expensive and beyond the reach of developing nations. In most of these countries sewage is discharged untreated into open drains and then on to rivers or open seas. Some countries including India and Thailand use a low-tech system of sewage treatment. The firststep of this system is the same as the hi-tech system. The liquid waste is then left in open pond for conversion of organic waste through photosynthesis. This low-tech system requires three times more space than the high-tech one. There are no programmes for solid-waste management. Most of solid waste is dumped on to open ground and used for landfil. Renewable resources like plastics, glass, metal etc. are however, recycled.

Index